ESSENTIAL MATHEMATICS

A Refresher Course for Business
and Social Studies

Clare Morris

and

Emmanuel Thanassoulis

MACMILLAN

© Clare Morris and Emmanuel Thanassoulis 1994

All rights reserved. No reproduction, copy or transmission of
this publication may be made without written permission.

No paragraph of this publication may be reproduced, copied or
transmitted save with written permission or in accordance with
the provisions of the Copyright, Designs and Patents Act 1988,
or under the terms of any licence permitting limited copying
issued by the Copyright Licensing Agency, 90 Tottenham Court
Road, London W1P 9HE.

Any person who does any unauthorised act in relation to this
publication may be liable to criminal prosecution and civil
claims for damages.

First published 1994 by
THE MACMILLAN PRESS LTD
Houndmills, Basingstoke, Hampshire RG21 2XS
and London
Companies and representatives
throughout the world

ISBN 0-333-57548-2

A catalogue record for this book is available
from the British Library.

Printed in Hong Kong

Essential
Mathematics

A19550

Contents

List of Figures   x

Preface   xii

1   ARITHMETIC OPERATIONS   1
1.1   **What is Arithmetic?**   1
1.2   **The Arithmetic of Whole Numbers**   2
     1.2.1   *Integers and the number line*   2
     1.2.2   *Addition and subtraction involving positive and negative integers*   2
     1.2.3   *Multiplication and division involving positive and negative integers*   4
     1.2.4   *Conventions about priority of arithmetic operations*   5
     1.2.5   *The role of zero*   6
     1.2.6   *Notation and terminology*   7
     1.2.7   *Writing large and small numbers*   8
1.3   **Fractions and their Arithmetic**   9
     1.3.1   *The place of fractions in the number system*   9
     1.3.2   *Fractional notation and terminology*   10
     1.3.3   *Addition and subtraction of fractions*   10
     1.3.4   *Multiplication of fractions*   12
     1.3.5   *Division of fractions*   12
1.4   **Decimals**   14
     1.4.1   *The decimal system*   14
     1.4.2   *Addition and subtraction of decimals*   15

Contents

- 1.4.3 *Multiplication and division of decimals by 10, 100 and so on*   16
- 1.4.4 *Multiplication of decimals*   16
- 1.4.5 *Division of decimals*   17
- 1.4.6 *Conversion between fractions and decimals*   18

1.5 Accuracy and Rounding   19
- 1.5.1 *Ways to specify the accuracy we require*   19
- 1.5.2 *How accurate do numbers need to be?*   21

1.6 Percentages   23
- 1.6.1 *What are percentages?*   23
- 1.6.2 *Calculation of percentages*   23
- 1.6.3 *Percentages with your calculator*   25
- 1.6.4 *Arithmetic with percentages*   26
- 1.6.5 *Percentage to decimal conversions*   28
- 1.6.6 *Use and misuse of percentages*   28

1.7 Some New Arithmetic Operations   30
- 1.7.1 *The summation sign*   30
- 1.7.2 *The modulus sign*   31
- 1.7.3 *The factorial sign*   32
- 1.7.4 *Inequalities*   32

2 ALGEBRAIC EXPRESSIONS   34

2.1 Introduction   34

2.2 Powers and Roots   38
- 2.2.1 *Introduction*   38
- 2.2.2 *Operations with powers*   41
- 2.2.3 *Extensions of the rules for operating with powers*   44
- 2.2.4 *Powers where the exponent is not a positive integer*   46
- 2.2.5 *Potential pitfalls to be avoided*   49

2.3 Using Brackets in Algebra   50
- 2.3.1 *Introduction*   50
- 2.3.2 *Removing brackets*   51

2.4 Factorising   55
- 2.4.1 *Factorising by taking common factors*   55
- 2.4.2 *Factorising by using identities*   56

Contents vii

3 FUNCTIONS AND GRAPHS 59
3.1 Introduction 59
3.2 The Basics of Graph-plotting 61
    3.2.1 *The Cartesian plane* 61
3.3 Linear Functions of a Single Variable and their Graphs 63
    3.3.1 *General form and its interpretations* 63
    3.3.2 *The graph of a linear function* 64
    3.3.3 *Features of the graph of a linear function of one variable* 65
    3.3.4 *More about slopes and intercepts* 67
    3.3.5 *Business applications of linear functions and their graphs* 71
    3.3.6 *A linear demand function* 71
    3.3.7 *A linear supply function* 74
3.4 Some Nonlinear Functions 78
    3.4.1 *Quadratic functions* 78
    3.4.2 *Exponential functions* 81
    3.4.3 *Inverse functions* 84
    3.4.4 *Logarithmic functions* 85
    3.4.5 *Semi-log graphs* 89
    3.4.6 *Log-log graphs* 92
    3.4.7 *Functions of the form* a/x^n 93

4 EQUATIONS 97
4.1 Introduction 97
4.2 Solving Linear Equations Involving One Variable 99
4.3 Solving Quadratic Equations with a Single Variable 104
4.4 Solving a Single Equation Involving more than One Variable 112
4.5 Simultaneous Equations 114
    4.5.1 *Introduction* 114
    4.5.2 *Solving two linear equations in two unknowns* 115
    4.5.3 *More complex cases of simultaneous equations* 122
4.6 Solving Linear Inequalities Involving a Single Variable 123

            4.6.1 *Introduction* 123
            4.6.2 *A method for solving linear inequalities with one variable* 124
            4.6.3 *Multiple inequalities* 126

5 INTRODUCTION TO CALCULUS 130
5.1 Introduction 130
5.2 Differentiation 131
   5.2.1 *Finding the slope of a curve* 131
   5.2.2 *Some rules for differentiation* 135
   5.2.3 *Higher order derivatives* 137
5.3 Optimising a Function 139
5.4 Integration 150
   5.4.1 *Background* 150
   5.4.2 *The process of integration* 151
   5.4.3 *Some uses of integration* 153

6 SOLVING PRACTICAL PROBLEMS WITH MATHEMATICS 157
6.1 Introduction 157
6.2 A Financial Problem 159
6.3 The General Problem-solving Process 163
6.4 A Depreciation Problem 163
6.5 A Stock-control Problem 166
6.6 Conclusion 171

7 SIMPLE STATISTICS 173
7.1 What is Statistics About? 173
7.2 How to Organise Statistical Data 174
   7.2.1 *Types of data* 174
   7.2.2 *Tabulating statistical data* 176
   7.2.3 *Frequency tables* 178
7.3 Statistical Diagrams 181
   7.3.1 *Diagrams for qualitative data* 181
   7.3.2 *Diagrams for quantitative data* 184
   7.3.3 *Other statistical diagrams* 190
7.4 Measures of Location and Dispersion 190
   7.4.1 *The need for summary measures* 190
   7.4.2 *Measuring location* 191
   7.4.3 *Measuring dispersion* 197
   7.4.4 *Choosing a measure* 202

Appendix 1 Some Uses of Integration   204
Appendix 2 Glossary of Technical Terms   214
Appendix 3 Solutions to Exercises   218
Index   248

List of Figures

1.1 The number line   3
3.1 The Cartesian plane   62
3.2 Graph of the function $6N + 1000$ 65
3.3 Illustrating slopes of linear functions   68
3.4 The link between slope and steepness   69
3.5 Illustrations of linear graphs   71
3.6 Supply and demand functions   73
3.7 Some typical parabolas   79
3.8 Graph of the function $y = x^2 - 4x + 2$ 80
3.9 Graph of the function $y = ab^x, b > 1$ 82
3.10 Graph of $y = e^x$ for $0 \leqslant x \leqslant 3$ 84
3.11 Graph of 10×0.4^x for $0 \leqslant x \leqslant 3$ 85
3.12 Graph of $y = \ln x$ for $0 \leqslant x \leqslant 3$ 87
3.13 Semi-log graph of interest data   91
3.14 Log-log graph showing $y = 10x^2$ 92
3.15 Graph of the function $y = a/x, a > 0$ 94
4.1 Wheat available after M months   99
4.2 Graph of $y = 5x - 15$ 100
4.3 Graph of $y = x^2 + x - 2$ 106
4.4 Graph of $y = x^2 + x + 2$ 107
4.5 Graph of $y = x^2 - 2x + 1$ 108
4.6 Graph of simultaneous equations   116
4.7 Inconsistent simultaneous equations   117
4.8 Plotting simultaneous inequalities on the number line   128
5.1 The slope changes along the curve   133
5.2 Computing the slope of $y = x^2$ at $x = 1$ 134
5.3 Optimising $R = (0.65 + P)(500 - 600P)$ 140
5.4 Graph of $R = (0.65 + P)(500 - 600P)$ and $dR/dP = 110 - 1200P$ 142

5.5	The local minimum of $y = 3x^2 - 12x$ is at $x = 2$	143
5.6	Local and absolute optimal points	145
5.7	A family of parallel curves	155
6.1	Using mathematics to solve problems	158
6.2	Graph of idealised stock cycle	169
7.1	Pie-chart of gender data	182
7.2	An overcrowded pie-chart	182
7.3	A bar-chart	184
7.4	Alternative form of bar-chart	185
7.5	A percentage bar-chart	186
7.6	Histogram of distance travelled	187
7.7	Histogram with unequal classes	187
7.8	Histogram wrongly plotted	188
7.9	Ogive of length of visit data	189
7.10	Calculating the median	196
A1.1	The definite integral as an area	207
A1.2	Summing definite integrals	209
A1.3	Areas below the x-axis must be computed separately	210
A1.4	A business application of definite integrals	211

Preface

This book originated in our experience of teaching preliminary mathematics courses for undergraduates about to embark on degrees in management, accounting and related disciplines, and for postgraduates about to start on MBA courses. Many of these students were nervous at the prospect of having to study a numerate discipline, whether called statistics, quantitative methods or whatever, as part of their intended course. For some, earlier experiences of studying mathematics had not been happy, and had left a legacy of confusion. Others had been perfectly competent mathematicians, but had not studied the subject for a number of years, and were anxious that they had become rusty. There was also a fear that certain topics which had not formed part of their previous experience – such as calculus – might be an essential prerequisite to their future course of study.

If you feel that the descriptions above apply to you, then this book is addressed to your needs. It is aimed at adults – so we hope you will find the tone not too reminiscent of a school-level text. It is designed for people who are primarily interested in professional subjects – business, management, accounting and so on – so there are practical examples drawn from those areas to illustrate and reinforce the mathematical principles. It is intended in the main for those working on their own; numerous worked examples, and practice exercises with full solutions, are therefore included.

The material of Chapters 1–4 is roughly that part of the GCSE O-level syllabus which could be described as arithmetic and algebra. For most readers, therefore, we envisage that much of this material will fall into the category of revision of things you once knew, rather than totally new material. Chapters 5–7 cover calculus, solving practical problems with mathematics, and elementary statistics, which are not always included in GCSE-level mathematics courses; in those chapters, we have imagined a reader to whom the concepts are completely new.

There are two points which we would like to make concerning the way you use this book. One is that no one ever learned mathematics by reading a book; the learning comes about through doing – through practising the techniques which have been described. So make sure that you do the exercises as you go along, and that if you disagree with one of the answers provided, you study the solution and ensure that you understand where you went astray before carrying on to the next section. Mathematics is to a large extent a linear subject – that is, one topic builds on another, so if you are confused about arithmetical operations you have little chance of getting to grips with algebra, and if algebra poses problems your grasp of calculus will probably be hazy. Thus the first five chapters of the book need to be read in sequence. If you find that you feel very comfortable with some of the material, you may wish to skim fairly quickly through the relevant text; we do, however, strongly recommend you to work through the exercises to verify whether you were right to feel confident!

The second point is chiefly a matter of attitude. Many people believe that mathematics is a rather mysterious and elusive subject, composed of abstractions, and accessible only to a minority of the population possessed of extraordinary minds. While it is true that there *are* areas of advanced mathematics to which this description would apply, it is also the case that elementary mathematics is extremely practical stuff, rooted in common sense, and useful for dealing with everyday problems in an efficient manner. That students studying the subject often do not have the opportunity to discover this is a sad reflection on those of us responsible for introducing them to mathematics. Our aim in writing this book has been to de-mystify the processes of elementary mathematics, and show just how useful they are in the context of business and management problems. If you, the reader, feel by the end of the book that it is really all rather simple stuff, we will have succeeded!

We are grateful to John Thanassoulis for checking the solutions to most of the exercises, and for drawing many of the diagrams. We would like to thank colleagues at Warwick Business School who have commented on earlier versions of some of this material produced for the Warwick Distance Learning MBA. We are also indebted to numerous students who, in attending the Preliminary

Mathematics Courses mentioned above, have unwittingly acted as guinea-pigs for our ideas. As always, any errors which remain are entirely our responsibility, and we will be happy to be informed of them.

CLARE MORRIS
EMMANUEL THANASSOULIS

1 Arithmetic Operations

1.1 What is Arithmetic?

At first sight the title of this section may strike you as pretty silly – after all, everyone knows what arithmetic is, don't they? It's something you used to have to do by hand, maybe even without pencil and paper (remember 'mental arithmetic'?), but which everyone now does on a calculator. It's concerned with the 'four rules' of addition, subtraction, multiplication and division, plus a few extras like percentages, and no one could describe it as interesting.

There is an element of truth in all this, but nevertheless it is worth giving a little more thought to the topic of arithmetic in general before we begin revising specific techniques.

Most of the 'maths' which is applied to business problems is, in fact, no more than arithmetic – albeit fairly complex arithmetic in some cases. Accounting, for example, was once defined to us by an accountant as 'the four rules – addition, subtraction, multiplication and division – and not very much of the last!' You only need to glance at the *Financial Times* to see how important percentages and their applications are – fundamental, for example, to the whole idea of investment and project appraisal. So it is clear that in order to feel comfortable in discussing these areas of business, you need to have a good grasp of basic arithmetical principles.

It is true that much of the drudgery of hand calculation can be dispensed with in these days of readily available software and cheap calculators. However, it is still necessary for the educated professional to acquire something of the 'feel for numbers' which really only comes from handling them. This does not mean being able to perform complex computations in your head – merely that you should be able to spot when the result of a calculation is very different from what you would expect, or when someone makes a statement which is clearly nonsensical in the light of the figures involved. You might argue that this skill is actually *more* rather than less necessary in the age of calculators and computers; we all know how very easy it is to transpose two digits when inputting a figure, and unless we have a good idea what we expect the result to be, such errors can go unnoticed, with disastrous and far-reaching consequences.

It is also true that the rather more advanced topics to be covered in later chapters, such as algebra, graphs and solution of equations, are all underpinned by arithmetic, so a good understanding of the basic rules and conventions is an essential starting point. The rest of this chapter will therefore help you to recall and refresh your arithmetical skills. We will not necessarily illustrate every point with a business application; the material of this chapter is really a foundation for more practical applications later in the book.

1.2 The Arithmetic of Whole Numbers

1.2.1 *Integers and the number line*

The mathematician's name for whole numbers such as 2, 3, 743 and −16 is *integers*. It can be helpful to think of the infinite set of such integers, both positive and negative, as strung out along a line, with positive numbers to the right of zero and negatives to the left, as in Figure 1.1. Then our arithmetic operations involve moving around on this line.

1.2.2 *Addition and subtraction involving positive and negative integers*

Most people are happy dealing with positive-integer arithmetic. However, negative quantities sometimes cause confusion. This is partly because the [−] sign is used in two different ways – as an operation, like addition or multiplication, and as a prefix to indicate a negative number. That's why you may have two separate keys on your calculator – a [−] key for the operation, and a [+/−] key to enable you to input a negative quantity.

FIGURE 1.1
The number line

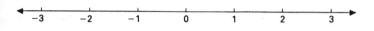

Actually, as we shall see, the two uses of the sign are very closely related – in fact, addition of a negative number is exactly equivalent to subtraction. For example,

$$7 + (-2) = 7 - 2 = 5.$$

Notice the use of the brackets here to prevent our having to write two signs next to each other, which looks odd. Later on we will encounter other uses for brackets; however, for those among you who have had some training in accounting, we should point out that the accounting convention by which (372) would be interpreted as -372 is *never* used in mathematics.

You may find it helpful to think of addition and subtraction in terms of moves along the line in Figure 1.1 – moves to the right in the case of addition, to the left for subtraction. Our calculation above of $7 + (-2)$ or $7 - 2$ involves starting at 7, then moving two to the left, so that we finish at 5. In the same way you can check that $-11 + 6 = -5$; $-7 - 2 = -9$ and so on. You may, of course, have learned alternative rules for carrying out this kind of arithmetic, such as 'in subtraction, put the sign of the larger number to the difference'. If you are happy with your old methods, by all means carry on using them – so long as they lead to correct answers!

Addition and subtraction are *inverse* operations for each other, in the sense that one 'undoes' the effect of the other. Subtracting 3 from a number will 'undo' the effect of adding 3 to it, and so on. We will encounter this inverse idea in several other contexts, and will find it useful when we come to solve equations in Chapter 4.

Probably the most familiar application of negative quantities arises in the context of costs, revenues, profits and losses. For example, if the cost of producing a tin of baked beans is 17p, and it can be sold for 23p, then the profit on the tin is obtained by subtracting 17p from 23p – that is, profit = $23 - 17 = 6$p. If, however, costs increase to 27p per tin, then if price remains unchanged the new profit will be 23 − 27 = −4p, a negative quantity signifying a loss of 4p per tin.

1.2.3 *Multiplication and division involving positive and negative integers*

Multiplication of positive integers presents no problems. However, a situation arises with division which we have not encountered before. When we add, subtract or multiply integers, we can be certain that the result will also be an integer, but the same does *not* apply to division. So we have to defer a general discussion of division of integers until later in the chapter, after we have covered fractions. For the moment, we concentrate on the case where the number we are dividing by 'goes into' the number to be divided a whole number of times. For example, because $3 \times 4 = 12$, we can say $12 \div 4 = 3$, so the result of the division is also an integer.

This example also shows that multiplication and division are inverse operations to each other in the same way as addition and subtraction: if we multiply a number by 4 and then divide by 4, we end up where we started.

To carry out multiplication and division involving negative quantities, you need to remember the following facts:

- **Multiplying or dividing two quantities with the *same* sign (+ or −) gives a *positive* result.**
- **Multiplying or dividing two quantities with *different* signs gives a *negative* result.**

It is possible to prove that these rules are essential if we are to have a consistent arithmetical system, so they are not purely arbitrary. However, you will probably find a practical justification simpler. As before, think of a negative quantity as a loss, so that a loss of £4 can be represented as −4. If we incur such a loss on three consecutive days, then the overall loss is clearly going to be £12, or −12. But this can also be thought of as $3 \times (-4)$; so we must interpret $3 \times (-4)$ as −12 – two unlike signs giving a negative result. Then since division is the inverse of multiplication, dividing −12 by −4 must get us back to 3 – two negative signs giving a positive result.

To see how the rules work, consider the following examples:

$6 \times (-7) = -42$ $\qquad$ $-8 \div 2 = -4$
$-5 \times -9 = 45$ $\qquad$ $-32 \div -4 = (+)8.$

The + in the last example is bracketed because we usually take it for granted – a number without a sign in front is assumed to be positive.

1.2.4 Conventions about priority of arithmetic operations

Thus far we have considered the individual arithmetic operations in isolation, but of course they are generally used in combination. The question then arises as to the order in which they should be carried out, since different orders could give different answers. For example, confronted with $23 - 2 \times 9$, do we proceed from left to right, first subtracting 2 from 23 to get 21, then multiplying this by 9 to get 189? Or do we do the 2×9 first, then subtract the resulting 18 from 23 to obtain 5? We need a convention to tell us which interpretation to use, otherwise the result would be considerable confusion!

The convention in general use is the following:

- **Multiplication and division take place before addition and subtraction.**
- **Multiplication and division have equal priority, as do addition and subtraction.**

So in the expression $23 - 2 \times 9$ considered above, multiplication takes place first, giving the answer 18. Had the expression been $23 + 2 - 9$, however, the two alternative orders

$$23 + 2 - 9 = 25 - 9 = 16$$

and

$$23 + 2 - 9 = 23 - 7 = 16$$

give the same result, since they involve only additions and subtractions. Likewise

$$3 \times 24 \times 8 = 72 \times 8 = 576,$$

and

$$3 \times 24 \times 8 = 3 \times 192 = 576,$$

because these expressions involve only multiplications.

Where division is involved, we need to be more careful: the expression 16/4/2, for example, cannot be evaluated as it stands; it could be interpreted as 4/2 = 2, or as 16/2 = 8. This is a case where the insertion of some brackets can make clear what order is required: (16/4)/2 or 16/(4/2). In fact, it is always possible to put in brackets to clarify the picture for yourself, even in a case where strictly speaking there is no ambiguity. So if you are happier writing 23 − (2 × 9) rather than remembering the priority rule, by all means do so.

This priority rule is built into many computing languages and spreadsheets, so you need to remember how 23 − 2 × 9 will be interpreted if, for example, you enter it as a formula in a spreadsheet cell. As we encounter other types of arithmetical operation later in the book, we will point out how they fit into the overall system of priorities. Meanwhile, here is a further example to help reinforce the idea.

An engineering assembly consists of three washers costing 2p each, two plates costing 15p each, and a bolt costing 40p. If we wish to work out the total cost of the assembly using a spreadsheet, how would we have to enter the calculation? The cost of the washers will be three times 2p, which has to be added to that of the plates – two times 15p – and the 40p for the bolt. The calculation thus can be written as 3 × 2 + 2 × 15 + 40, the multiplications conventionally being carried out first. However, if you wanted to make it absolutely clear you could write (3 × 2) + (2 × 15) + 40. (Of course, you would probably do this very simple calculation in your head without even thinking about it – but later on when we want to use algebraic symbols to replace the numbers in this kind of calculation, you will need to be more aware of the rules.)

1.2.5 *The role of zero*

The number zero is rather special – in fact, the point in history at which Islamic scholars first invented a symbol for zero is regarded as a landmark in the development of mathematics. The effects of this special status in the light of our discussion thus far are threefold:

(a) **Adding zero to a number, or subtracting it from a number, leaves the number unchanged**: so $11 + 0 = 11$, and $18 - 0 = 18$.
(b) **Multiplying any quantity by zero gives an answer of zero.** You could interpret this in a colloquial way as 'any number of nothings is still nothing'.
(c) **Division by zero is not possible within the ordinary arithmetic system**. If you have a good calculator you can confirm this by trying to carry out an operation such as $20 \div 0$ – an error message should be displayed. It is sometimes said that division by zero gives an answer of infinity, but this statement has to be carefully interpreted: certainly if we divide 20, say, by smaller and smaller quantities, the results become bigger and bigger (try dividing by 0.1, 0.01, 0.0001 and so on), so that as the number being divided by approaches zero, the answer approaches infinity. However, from a practical point of view this does not mean very much, so if you find yourself, in the middle of a computation, attempting a division by zero, you have probably made a mistake somewhere along the line.

1.2.6 *Notation and terminology*

The notation + and − for addition and subtraction respectively is standard. Multiplication is generally denoted by × in print and on calculators, but to avoid confusion with the letter x, most programming languages use ∗ instead. Division shows the greatest diversity of notation; at school we learn to write ÷, but the oblique line / is preferred in computing languages, so that $15/5 = 3$. And as we will see in the next section, the notation $\frac{15}{5}$, which we more often associate with fractions, can also be interpreted as a division.

The various numbers involved in calculations have been given technical names, as follows:

- The result of adding a set of numbers is called their *sum*; thus 12 is the sum of 3, 4 and 5.
- The result of a subtraction is a *difference*, so that 2 is the difference between 12 and 10.
- When two numbers are multiplied, we refer to their *product*; 10

is the product of 5 and 2.
- In a division, the number we are dividing by is the *divisor*, that being divided into the *dividend*, while the result of the division is a *quotient*. Thus in 12/6 = 2: 12 is the dividend, 6 the divisor and 2 the quotient.

1.2.7 Writing large and small numbers

In a business context we often need to deal with very large numbers – turnovers in millions or even billions of dollars, production of thousands of items per day, and so on. It is often simpler to quote a turnover as, say, £82m rather than to quote it to the last penny (a point to which we shall be returning in section 1.5), but then you need to know just how many zeros are involved should you need to use the figure in a calculation. Since 1 million = 1 000 000, we can express £82m as £82 × 1 000 000 = £82 000 000. In the same way, if we're told that daily production of widgets in a factory is sixteen thousand, that can be written as 16 × 1000 = 16 000.

Billions are more problematic – there's a difference between the British definition (one million million or 1 000 000 000 000) and the American version (one thousand million or 1 000 000 000). So you need to find out which is meant before trying to write such numbers. In fact, it is clearly inefficient to use such large numbers of zeros, which for most people become rather mind-boggling. This is why many calculators and computer programmes use an alternative method for writing very large numbers, often called *scientific* or *exponent* notation. You have probably seen this on calculator displays – a number written in the form 8E5, for example. A full discussion of how it works will have to wait until we have covered the concept of *powers* in Chapter 2, but essentially the 5 after the E tells us that we have to multiply the 8 by a 1 with five 0's. So 8E5 = 8 × 100 000 = 800 000. In the same way, 6 million could be written 6E6, and so on.

Exercises 1.1

1. Evaluate the following expressions, without using a calculator:

(a) 8 − 13;
(b) −11 + 5;
(c) 25 ÷ (−5);
(d) (−7) × (−8).
2. What is the sum of 12, −23 and 8?
3. What is the product of −18 and 4?
4. A company has four divisions, which in a particular year make respectively profits of £23m and £31m, and losses of £6m and £200 000. What is the overall profit or loss of the company for the year?
5. One of a stockbroker's clients has a portfolio consisting of 120 shares in company A, 250 in B, and 75 in C. If company A this year is paying a dividend of 15p per share, B is paying only 4p, and C is paying no dividend at all, how much dividend in total can the client expect to receive?

1.3 Fractions and their Arithmetic

1.3.1 *The place of fractions in the number system*

When discussing the number line in the last section, we talked as though it consisted of a series of integers with gaps in between. In fact, of course, that is not so – the gaps are occupied by fractional numbers (and others, as we shall see later). We shall assume that you have an understanding of what is meant by a fraction such as 1/4, 2/3, etc., so will concentrate on how they are used in arithmetic operations.

Before that, however, it may be helpful to consider why an understanding of fractional arithmetic is needed, even though a fraction can generally be converted to a decimal and handled via a calculator. A partial answer lies in the fact that certain fractions cannot be expressed precisely as decimals; you probably remember, and we will demonstrate later, that $1/9 = 0.1111\ldots$, a recurring decimal which never terminates. Much more accurate, then, to quote the fraction than the decimal.

It is also often easier for people to interpret fractions; being told 'about a quarter of our customers take more than 3 months to settle their accounts' is more meaningful than 'the proportion of customers taking more than 3 months is 0.253'. This is especially true when dealing with the assessment of probabilities or chances; more managers would be happy with the statement 'the chance that we'll get this contract is about 1 in 5 (or 1/5)' than 'the chance of getting this contract is 0.21'. Not for nothing do bookmakers customarily work with 'odds'!

Given, then, that we want to be able to use fractions in this way, we need at least to know the rules for operating with them.

1.3.2 Fractional notation and terminology

The most familiar way of writing fractions is $\frac{1}{2}$, $\frac{3}{7}$, etc., but we can also write 1/2, 3/7, etc. Since this is identical with one of the notations we used for division in section 1.2, it is clear that a fraction may also be viewed as a division – a point to which we shall return when we discuss decimals.

There is no reason why we should not have a fraction which is greater than 1 – for example, $\frac{11}{8}$. Sometimes the whole number part of such a fraction is separated out, and the result written as a 'mixed fraction', $1\frac{3}{8}$, though this is not very common in mathematics.

The 'number on the top' of a fraction is called its *numerator*, and that 'on the bottom' its *denominator*. Many different combinations of numerator and denominator can be used to represent the same fraction, as you will see if you consider $\frac{1}{2}$, $\frac{3}{6}$, $\frac{8}{16}$, etc. In each case, the fraction represents the same proportion of the whole; whether we take three-sixths of a total, or eight-sixteenths, we end up with the same amount – namely, one half.

The equivalence of these fractions is the basis for the process known as *cancelling* – dividing the numerator and denominator of a fraction by the same number. For instance, dividing both top and bottom of $\frac{3}{6}$ by 3 gives $\frac{1}{2}$, while with $\frac{8}{16}$ we can divide by 8. A fraction which cannot be cancelled – such as $\frac{1}{2}$ – is said to be *in its lowest terms*, and since simplicity is a virtue we prefer, on the whole, to express fractions in their lowest terms wherever possible.

1.3.3 Addition and subtraction of fractions

Most people, when fractions are mentioned, find the words 'common denominator' floating to the surface of their minds. This is because fractions which already have the same denominator can be added directly; for example, $\frac{2}{5} + \frac{1}{5} = \frac{3}{5}$. However, fractions with different denominators cannot be directly added in this way – trying to add $\frac{2}{7}$ and $\frac{3}{4}$ is rather like trying to add amounts of money expressed in different currencies. If we want to find the cost of building the Channel Tunnel by combining the costs incurred in pounds sterling and in French francs, we need to express both sets of costs in a common unit first – perhaps converting both to Ecu, or converting the francs to sterling.

What, then, is a sensible common basis to which we can reduce $\frac{2}{7}$ and $\frac{3}{4}$? One possible answer (not, as we shall see, the only one) is to express both fractions in a form with 28 as the denominator – 28 being the product of 7 and 4. To express $\frac{2}{7}$ in twenty-eighths, we perform a sort of reverse version of the cancellation operation mentioned above – that is, we *multiply* both numerator and denominator of the fraction by 4, obtaining $\frac{8}{28}$. This does not, of course, change the value of the fraction. For $\frac{3}{4}$, we multiply top and bottom by 7, getting $\frac{21}{28}$. Now, since both fractions are expressed in the same 'unit' of twenty-eighths, we can add their numerators to get $\frac{8}{28} + \frac{21}{28} = (8 + 21)/28 = \frac{29}{28}$.

The trick we used in this example to find the common denominator was to multiply the denominators of the two individual fractions. This will always work, but it does not always give the most efficient result. Consider $\frac{5}{6} - \frac{3}{4}$ (subtraction requires a common denominator also). Here we could use 24 as the common denominator, getting $\frac{5}{6} - \frac{3}{4} = \frac{20}{24} - \frac{18}{24} = \frac{2}{24}$. But the fact that $\frac{2}{24}$ can be cancelled to $\frac{1}{12}$ suggests that we have to some extent gone round in a circle – first multiplying tops and bottoms of fractions, then dividing them. The reason is that 12 also provides a common denominator for 6 and 4, and being smaller it's a more efficient one. So we can say $\frac{5}{6} - \frac{3}{4} = \frac{10}{12} - \frac{9}{12} = \frac{1}{12}$ – the same answer as before, but in fewer steps. The number 12 is called the *lowest common denominator* for a calculation involving sixths and fourths, because it is the smallest number into which both 6 and 4 will divide evenly.

Let us summarise the position on addition and subtraction:

- **To add or subtract fractions, express them in terms of their lowest common denominator – that is, the smallest number into which all the individual denominators will divide evenly.**

Here is another example to illustrate the idea:

$$\frac{4}{15} - \frac{2}{3} + \frac{1}{6} = \frac{(8 - 20 + 5)}{30} = -\frac{7}{30}$$

1.3.4 Multiplication of fractions

This is the easiest of all fractional operations, since we simply *multiply the numerators and denominators*, and then do any necessary cancelling:

$$\frac{3}{5} \times \frac{8}{9} = \frac{24}{45} = \frac{8}{15} \text{ (cancelling by 3)},$$

$$\frac{11}{4} \times \frac{7}{8} = \frac{77}{32} \text{ and so on.}$$

Alternatively, the cancelling can be done first, which has the advantage of not leading to such large numbers:

$$\frac{3}{5} \times \frac{8}{9} = \frac{(3 \times 8)}{(5 \times 9)} = \frac{(1 \times 8)}{(5 \times 3)} = \frac{8}{15} \text{ as before.}$$

One point which sometimes causes confusion is the multiplication of a fraction by an integer, as in $8 \times \frac{7}{12}$. Any integer can in fact be regarded as a fraction whose denominator is 1, so that 8 is $\frac{8}{1}$. The multiplication then gives $\frac{8}{1} \times \frac{7}{12} = \frac{56}{12} = \frac{14}{3}$ in the usual way.

1.3.5 Division of fractions

It is probably easiest here to give the 'rule' first, and then see how it works. Suppose we want to divide $\frac{3}{5}$ by $\frac{1}{5}$ – that is, to find $\frac{3}{5} \div \frac{1}{5}$. The rule is usually expressed concisely as:

- **Invert and multiply.**

Or, more lengthily:

- **Invert the fraction you are dividing by and multiply it by the one you are dividing into.**

So we invert the divisor – the $\frac{1}{5}$ in this case – and multiply the $\frac{3}{5}$ by the result, obtaining:

$$\frac{3}{5} \div \frac{1}{5} = \frac{3}{5} \times \frac{5}{1} = \frac{15}{5} = 3.$$

You can see that this is correct, since there are three × one-fifth in three-fifths. Another example:
$$\frac{3}{8} \div \frac{2}{5} = \frac{3}{8} \times \frac{5}{2} = \frac{15}{16}.$$
A particularly useful consequence of this rule is that
$$12 \div 3 = 12 \div \frac{3}{1} = 12 \times \frac{1}{3},$$
so that dividing by 3 and multiplying by $\frac{1}{3}$ are seen to be equivalent. A similar result applies for any integer division.

Exercises 1.2

1. Evaluate the following expressions:
   (a) $\frac{2}{3} + \frac{2}{5}$;
   (b) $\frac{4}{9} - \frac{1}{6}$;
   (c) $\frac{11}{12} \times \frac{3}{7}$;
   (d) $\frac{1}{15} \div \frac{5}{6}$.
2. What is the sum of $\frac{8}{9}$ and $\frac{2}{3}$?
3. Find the product of $\frac{3}{4}$ and $\frac{8}{9}$.
4. Divide 6 by $\frac{1}{4}$.
5. What is one-twelfth of one-and-a-half?
6. Market research indicates that two-thirds of a firm's customers prefer to pay by credit card; one-eighth use cheques, and the remainder pay by cash. If the firm has a base of 18 000 customers, how many cash payers can it expect?
7. Many British higher education institutions are now using the Credit Accumulation and Transfer (CATS) system, whereby a student can accumulate 120 CATS points for taking a full-time one-year course (and pro rata for fractions of a year). Thus a year could be split into 4 subjects, each carrying $\frac{1}{4}$ of the points (i.e. 30); or into 5, each carrying 24 points. How many different splits into equally-weighted subjects are possible while retaining whole numbers of credit points? Does this suggest a possible reason why 120, and not the more obvious 100, has been chosen?
8. The Infant Mortality Rate (IMR) is an important measure of the standard of living in a country. In the UK in 1953, 28 out

of every 1000 live babies born died within their first year of life. What is this rate as a fraction in its lowest terms? At present (1993) the corresponding figure is 8 in 1000. Express this as a fraction on the same denominator as your figure for 1953. How much greater was the rate forty years ago?

1.4 Decimals

1.4.1 *The decimal system*

We are so accustomed to our decimal, or ten-based, number system that we tend to take it for granted, but if we are to understand how 'decimals' in the usual sense – that is, decimal fractions – behave, then we first need to think a little about our method of writing numbers.

We have what is known as a *place-value system*; the place of a digit within the entire number specifies the value which is to be assigned to it. So, for example, the 2 in the number 234 has a different value from that in the number 32; in the first case, the 2 actually represents two hundreds, whereas in the second it is two units. In our system, the value of the digit increases with its place in what are called *powers* of 10 – 100, 1000 and so on (we will be looking at the idea of a power more closely in Chapter 2). So in an integer such as 234, the digit furthest to the right (4 in this case) denotes the number of units, the second from right – 3 – the number of tens, and the third from right – 2 – the number of hundreds. Thus the entire number is to be interpreted as $4 \times 1 + 3 \times 10 + 2 \times 100$.

This is not by any means the only possible system. You may have come across binary, octal or hexadecimal systems used in computing, which work on multiples of 2, 8 and 16 respectively, and if this were a book on the history of mathematics we could discuss the way other cultures have chosen to write their numbers (like the rather clumsy Roman numeral system). But we are more interested in the extension of the decimal system to fractions.

In fact, just as we can add digits to the left to represent higher multiples of 10, so we can add them to the right to represent fractions based on ten – tenths, hundredths and so on. However, in order to keep track of where we are, we need a way of marking the position of the 'units' digit – and that is exactly what the decimal point does. So when we write 234.96, the 2, 3 and 4 are interpreted as already explained, the decimal point [.] comes immediately after the units digit, and then the 9 denotes $9 \times \frac{1}{10}$ and the 6, $6 \times \frac{1}{100}$.

If we want to write a very small number, perhaps one which does not involve tenths or hundredths, then we have to fill up the empty places with zeros: thus 0.0056 means $\frac{5}{1000} + \frac{6}{10000} = \frac{56}{10000}$. (You don't have to put the initial zero here; the number could be written as .0056, but it's probably better to include the zero before the decimal point, to avoid the risk of the decimal point being overlooked.) A shorthand way of describing the transition from decimal 0.0056 to fraction $\frac{56}{10000}$ is to say that the denominator contains a 1 for the point and a zero for every figure after the point'.

You can see from this that any integer can be written as a decimal with one or more zeros after the point; 43 = 43.0 (i.e. 4 tens, 3 ones, and no tenths) and so on. In theory you could add as many zeros after the point as you liked, without affecting the value of the number. In practice there are conventions about the way we interpret figures written in this way, which will be discussed later, in Section 1.5, 'Accuracy and Rounding', but for the moment you do not need to worry about them.

1.4.2 Addition and subtraction of decimals

There is really nothing new to learn about these operations: they work in exactly the same way as addition and subtraction of integers. The only point you need to remember is that the decimal points should be lined up so that all the tenths, hundredths and so on are in the same vertical column.

EXAMPLE

```
   1.3
   2.97
   0.04
  10.5
  ─────
  14.81
```

1.4.3 Multiplication and division of decimals by 10, 100 and so on

Before looking at general multiplication and division of decimals, let us consider a rather easy special case: multiplication or division of a decimal by a power of ten (100, 1000, etc.). The place-value system underlying the decimal number system means that these operations are especially simple. The rule for multiplying by powers of ten is:

- **To multiply a decimal by 10, move the point one place to the right; to multiply by 100, move the point two places to the right, etc.**

Thus $5.8 \times 10 = 58$; $5.8 \times 100 = 580$ (because 5.8 could be written 5.80) and so on. What we are doing here is implicitly using the fact that 5.8 means $5 \times 1 + 8 \times 1/10$, so $5.8 \times 10 = 5 \times 10 + 8 \times 1$ – all the powers of 10 have gone up by one, which is equivalent to moving the point one step to the right in our place-value system. However, you do not need to think through this argument every time in order to use the rule!

To divide, as you might expect, we move the point in the opposite direction:

- **To divide a decimal by 10, move the point one place to the left; to divide by 100, move the point two places to the left, etc.**

So $18/10 = 1.8$ (remember that 18 is really 18.0), and $24.6/100 = 0.246$.

1.4.4 Multiplication of decimals

To multiply two decimals together, apply the following procedure. Forget about the decimal points, and just multiply the numbers; for example, to multiply 0.03 by 0.008, start with $3 \times 8 = 24$. Think of this as 24.00. . . . Then find where the decimal point in the answer should be by using this rule:

- **To multiply two decimals together, move the decimal point to the left a number of places equal to the total of the numbers of figures after the point in the two separate numbers being multiplied.**

In the example above, there are two figures after the point in 0.03 and three in 0.008, giving a total of five. So applying the rule means we move the point in 24.000 five places to the left, putting in zeros as necessary, to get 0.00024.

Like many arithmetical procedures, this is more complicated to explain than it is to carry out. To see why the rule operates as it does, think about the fractional equivalent of the decimals. In our example above, for instance, $0.03 \times 0.008 = \frac{3}{100} \times \frac{8}{1000} = \frac{24}{100000} = 0.00024$, which is what we got by applying the rules.

The process works in much the same way if we have figures greater than 1 to deal with. To find 3.2×25, begin by multiplying 32 by 25, giving 800. Then count the decimal places involved: 3.2 has one figure after the point, and 25 has none, so the total is 1. We therefore need to move the point in the answer one place to the left, so we end up with 80.0, or simply 80.

1.4.5 Division of decimals

We start by examining the division of a decimal by an integer; let us take as an example 0.12/3. In fractional terms this is $\frac{12}{100}$ divided by $\frac{3}{1}$, which the rules for dividing fractions tell us is $\frac{12}{100} \times \frac{1}{3} = \frac{4}{100} = 0.04$. In other words, the answer we get is exactly that which would result from dividing as we do with integers, remembering to put in the decimal point when we encounter it. Another example would be $\frac{3.75}{5} = 0.75$.

Now suppose we want to divide any two decimal numbers; an example might be $\frac{4.148}{0.04}$. It's easy to turn this into division by an integer if we remember the point made earlier that multiplying top and bottom of a fraction by the same number does not alter its value. In our example, if we multiply both top and bottom by 100, we get $\frac{414.8}{4}$ – in other words, we now have a whole number on the bottom. So we can now proceed as in the last paragraph, to get 103.7 as the answer.

What we have done here can be summed up in **the rule for division of decimals**:

Multiply both divisor and dividend by a power of ten (i.e. 10, 100, 1000, etc.) big enough to remove decimal places from the divisor. Then carry out the process of dividing by a whole number, as described above, putting in the decimal point when you come to it.

1.4.6 *Conversion between fractions and decimals*

Fractions and decimals can be used interchangeably, since as we now know a decimal is just a special sort of fraction. To turn a fraction into a decimal, the process is just that of decimal division, as described in the last paragraph, except that there are no explicit decimal points in numerator or denominator. For instance, if we want $\frac{1}{20}$ as a decimal, we carry out the division $1.00\ldots/20 = 0.05$.

There are cases where this division never actually comes to an end, no matter how long we carry on dividing, so that we end up with what's called a *recurring decimal*. An obvious case is $\frac{1}{3}$, which if divided out becomes $0.3333\ldots$ – there is always another remainder of 1 to divide into. Sometimes not just one but a whole group of figures recurs – see what happens if you try to convert $\frac{1}{11}$ to a decimal.

To save time, most people store the decimal versions of some simple fractions in their heads: $\frac{1}{2} = 0.5$, $\frac{1}{4} = 0.25$, $\frac{1}{5} = 0.2$, etc.

It is worth noting that we have now dealt with the problem encountered in section 1.2: how to divide two integers when one doesn't 'go' an even number of times into the other – for example, $\frac{7}{3}$? You can see that this is exactly the same question as converting the fraction $\frac{7}{3}$ to a decimal, and gives $2.333\ldots$.

Conversion of decimals to fractions is even easier: just remember that a decimal is a fraction with 10, 100, or some other power of 10 as its denominator, and then cancel if necessary. Thus $0.23 = \frac{23}{100}$; $0.46 = \frac{46}{100} = \frac{23}{50}$ (cancelling by 2) and so on.

Exercises 1.3

1. Perform the operations:
   (a) $0.24 + 3.791$;
   (b) $4.6 - 8.92$;
   (c) $2/0.1$;
   (d) 10.4×20;
   (e) 16.71×0.05;
   (f) $173.42/0.12$.
2. What is $\frac{8}{9}$ as a decimal?
3. What is 0.325 as a fraction in its lowest terms?
4. How many Norwegian kroner will I receive in exchange for a sum of £102.19, on a day when the rate of exchange is 10.13 kroner to the pound?

5. If the dollars to sterling exchange rate is 1.43 dollars to the pound, and a company wishes to buy a machine costing $13,492, what will be the equivalent price in £?

1.5 Accuracy and Rounding

1.5.1 *Ways to specify the accuracy we require*

You have already encountered some numbers which cannot ever be written down with complete accuracy in decimal form, because their decimal expression never terminates – remember trying to express $\frac{1}{11}$ as a decimal? The numbers we have met so far were recurring decimals, but there are actually some numbers which neither recur nor terminate – they simply go on as a string of digits in no predictable order forever. You probably remember π (the ratio of the diameter of a circle to its circumference) as a number of this kind – at school you may have written it as 3.142, or even $\frac{22}{7}$, but these are only approximations. Another such number which you will meet in later chapters is e, called the *base of natural logarithms*. This number, which crops up in all sorts of mathematical contexts, is approximately equal to 2.71828, but as with π its full decimal expression neither recurs nor terminates.

Even when we *can* express a number accurately in a finite number of figures, we may not want to. Perhaps we just want to get a general idea of how big the number is; perhaps there is some doubt about the accuracy of all the decimal places. There is clearly a need for a method of specifying the level of accuracy we require in a number. We will first look at three ways of doing this, and then consider the related question of how to decide what is an adequate level of accuracy in a given context.

a) *Decimal places* If the numbers we are dealing with involve decimals, this is a simple way of specifying their accuracy. So we might say that we wish calculations to be done 'correct to two decimal places'. Then a number such as 4.983 would become 4.98, because 4.983 is closer to 4.980 than to 4.990. On the other hand, 4.987 would become 4.99. The usual rule is that digits 1, 2, 3, 4 are taken to the next lower number, while 5, 6, 7, 8, 9 are taken upwards (zeros, of course, do not need changing). So 4.981, 4.982, 4.983, and 4.984 would all become 4.98 to two decimal places, while 4.985, 4.986, 4.987, 4.988 and 4.989 would go up to 4.99.

This brings us back to a point we mentioned earlier. Adding zeros to the end of a decimal figure implies accuracy to a certain number of decimal places: 4.630, for example, suggests that the figure is correct to 3 places, while 4.6300 would imply 4-figure accuracy, and 4.63 would only guarantee two figures. (The way we interpret sums of money in sterling illustrates this: if we see the figure £2.40, we understand it to mean two pounds and 40 pence, not 41 pence or 39 pence – we apply an implicit two-decimal-place accuracy to the number.) So adding unnecessary zeros to the ends of figures is not a good idea, unless you really mean them to be interpreted in this way.

(b) Rounding An alternative method, particularly useful when we are dealing with large numbers, is to ask for them rounded 'to the nearest hundred', 'to the nearest thousand', and so on. The phrase is self-explanatory: 2300 to the nearest thousand is 2000; 27.9 to the nearest whole number is 28, etc.

As with specifying decimal places, we have to decide what to do about numbers which are poised exactly at the mid-point of the relevant range. If we are trying to find 850 to the nearest hundred for example, do we go to 800 or 900? Again, the usual convention is to take these 'mid-point' values up, so 850 would become 900 and 750 would become 800.

There's actually a small upward bias in this system; for example if we are rounding to the nearest hundred, then all figures from $x01$ to $x49$ are rounded downwards, while those from $x50$ to $x99$ go upwards. This means that 50 digits get rounded up, but only 49 down – a fact which may seem trivial, but could lead to major errors if, say, the method were being used over a long period of time on calculations involving large amounts of money. In such a case, an alternative system which rounds the halfway points to the nearest *even* number might be preferred. Under this system, 750 would be rounded to 800 as before, but 850 would also round *down* to 800. This results in some mid-points being rounded upward and some downward. In the long term the numbers of upward and downward roundings will balance out.

Generally speaking we would recommend you to stick with the system you are used to using – the most important point is to be consistent.

(c) Significant figures This is one of those ideas which are much easier to understand from an example. In the number 604 400 the first four digits (the 6, 0, 4 and 4) are significant in the sense that they convey information about the value of the number; the zeroes on the end are not significant in this way, since they are only there to indicate place values. Again, for 0.00406, the 4, the last 0 and the 6 are significant, the initial two zeros are not. So we can say

- **Zeros are significant figures only when preceded and followed by non-zero digits**.

With this definition understood, we can specify how many significant figures we require instead of specifying the number of decimal places or the level of rounding. So 0.00406 to two significant figures is 0.0041; 604 400 to three significant figures is 604 000, and so on.

1.5.2 *How accurate do numbers need to be?*

Calculators, and still more computers, often give answers to a great many decimal places. The question then arises as to just how many of these one should actually use. Unfortunately we cannot give a simple definite answer to this question, since the number of figures to be retained really depends on two factors:

(a) How are you going to be using the figures? If you just want to convey an idea of the general size of a quantity, it can be a distraction to quote very detailed figures. If, for example, you are describing the floor area of a new office block for advertising purposes, it is pointless to give it as 1246.97 square metres – a statement such as 'nearly 1250 square metres' would be much more sensible. In fact, it would probably be hard to determine the area to two-decimal-place accuracy, so at least the last figure may be rather suspect – an example of 'spurious accuracy'.

If, on the other hand, you are quoting an interest rate – say, the APR (Annual Percentage Rate) of a loan – then the third or fourth decimal place might be crucial. The difference between 12.5% and

12.53% looks trivial, but by the time it has been multiplied by a sum borrowed of, say, £10 000, the difference is £30 – quite a substantial amount.

These examples should make it clear that you need to use common sense in developing a sound instinct for the number of figures to retain in a given situation.

(b) What was the accuracy of the figures which went into the calculation? It is no good quoting an answer to four decimal places if some of the figures involved in its calculation were only accurate to the nearest whole number. This becomes clear if we look at an example.

Suppose we are looking at the weekly production of three lines in a factory, for items which are manufactured in very large quantities. The supervisor of line A quotes the figure to the nearest 100 as 15 600; the supervisor of line B gives her total as 18 000, rounded to the nearest 1000; while line C's supervisor, a stickler for precision, claims that his production was 16 643.

Straight addition of these three figures, taken at their face value, would give 50 243. However, the 15 600 could really be as high as 15 649 or as low as 15 550, while the 18 000 could be anywhere in the range 17 500 to 18 449. Thus the total (assuming that we believe the 16 643 to be absolutely accurate) might be anything from 48 693 to 50 741 – a range of more than 2000. Quoting the result as 50 243 is therefore misleading; we can either round all the figures to the level of the least accurate, getting 16 000 + 18 000 + 17 000 = 51 000 – or give clearer instructions to the supervisors as to how they should deal with the figures.

You need to bear this consideration in mind when looking at other people's figures. For example, are the last few digits *really* credible? Even if they were accurate when collected, are they still correct now? (Think of the ten-yearly Census – there will have been many births and deaths between collection and publication of the population figures).

Exercises 1.4

1. Find 8.9266, 24.36879, 0.004572, to 3 decimal places.
2. What are 10 170, 0.03064, 2.6666 to 3 significant figures?

3. Weekly takings of three branches of a retail store are quoted as £24 279 (correct to the nearest pound); £18 730 (to the nearest ten pounds); and £19 350 (to the nearest £50). How accurately can you quote the total takings?

1.6 Percentages

1.6.1 *What are percentages?*

We have already seen that decimals are really a type of fraction, based on powers of 10. Percentages are just another sort of fraction, this time based on a standard denominator of 100. So when we write 32%, what we mean is $\frac{32}{100}$ – in fact, the % sign is what remains of the /100 written very fast!

The great strength of using percentages is seen when we want to compare awkward proportions. To most of us, the statement, 'Last year we offered jobs to 47 out of 93 applicants; this year we have taken 38 out of 80' is not easy to interpret. If we are good at mental arithmetic, we may notice that the proportion offered jobs has fallen a bit, since 47 is just over half of 93, while 38 is just under half of 80; but any deduction more precise than this is difficult, because of the fact that the totals are different, and neither of them are particularly 'nice' numbers.

If, however, we are told that the proportion of applicants being offered jobs was 51% last year, and this year it is 48%, we get a much better idea of the change. This is simply because we are now comparing figures on a common base – and a base which, being 100, is easier to visualise.

In spite of the fact that percentages are a simple idea, there's a lot of confusion about their calculation and use. The rest of this section will attempt to clarify these points; if we seem to be spending a long time over the topic, that is because percentages are very widely used – and frequently *mis*used.

1.6.2 *Calculation of percentages*

It is important to realise from the outset that we use percentages in two quite different situations.

(a) Finding **a** *as a percentage of* **b**

The first is where we want to calculate one quantity as a percentage of another. We had an example of this above: if we offer jobs to 38 out of 80 applicants, what percentage does this constitute? In effect, we want to convert the fraction $\frac{38}{80}$ to a denominator of 100, so we are saying 'What number out of 100 constitutes the same proportion as 38 out of 80?' – that is, we need to solve the equation $\frac{x}{100} = \frac{38}{80}$.

We shall not be revising solution of equations until Chapter 4, but this is such a simple equation that you can probably see that the solution is: $x = 38/80 \times 100 = 47.5$. So 38 is 47.5% of 80. What we have done here can be generalised to give a formula for calculating any figure as a percentage of another:

- **To express a as a percentage of b, calculate $(a/b) \times 100$.**

Thus 6 as a percentage of 24 is $(6/24) \times 100 = 25\%$; 9 as a percentage of 27 is $(9/27) \times 100 = 33\%$ to the nearest whole per cent, and so on.

(b) Finding **a**% *of a quantity* **b**

Many percentage calculations are the reverse of the one in the last paragraph – for example, we know what a bill is without VAT, and we want to find 17.5% of that bill (17.5% being the VAT rate in the UK at the time of writing). If the bill is £80, for example, then we need 17.5% of £80. Recalling that the % is a fraction based on 100, we can calculate $17.5/100 \times £80 = £14$. The general rule based on this is:

- **To find $a\%$ of b, calculate $(a/100) \times b$.**

To avoid confusion, be very clear about whether you are using a % in the context of (a) or (b) above.

It follows from these rules that merely talking about a percentage without saying 'percentage of *what*?' does not make sense. Sometimes you can answer this question from the context – it is pretty obvious that if a restaurant menu says '15% service charge added' it means 15% of the total bill. However, if there is any doubt, then the base for the percentage needs to be made clear: for example, 'the cost of living fell this month by 3% compared with this time last year'; or 'the proportion of reject items constitutes 4% of the total items produced by the factory'.

A familiar example makes clear the need to be precise about the base of the percentage. If a bill including 17.5% VAT is £200, what was the bill before the addition of VAT? If your answer is £165, you have made a popular mistake. To see why, try adding 17.5% to £165 – you'll find the result is £193.88, NOT £200 as it should be. The problem is that you have not asked the question '% of *what*?' – you have assumed that the 17.5% is 17.5% of the final £200, whereas, of course it is 17.5% of the amount *before* the VAT was added. The true answer is in fact £170.21, as you can check by adding 17.5% to £170.21. You should get £200, give or take the odd penny.

How did we arrive at the answer of £170.21? The answer really requires an understanding of the solution of equations, which we will not formally be covering until Chapter 4. However, for those of you who would like to see the argument at this point, what we have said is that the £200 is in fact 117.5% of the original bill. If we use £y to denote the amount of the original bill, then we can express this fact as an equation by saying 117.5% of y = 200. Recalling that a percentage is a fraction with a denominator of 100, this can be rewritten as $117.5y/100 = 200$, so that $y = 170.21$.

1.6.3 *Percentages with your calculator*

The [%] key on a calculator does away with the need to enter the ×100 and ÷ 100 in percentage calculations. Not all calculators work in the same way; for full details of your own, be sure to read the instruction manual. However, it is worth trying the following sequence, which works on many simple calculators: to find a as a percentage of b, enter a divide b, but press the % key instead of the =. The ×100 is then carried out automatically. So, for instance, to calculate 6 as a percentage of 24, key in 6 ÷ 24 %. The answer, as we know from our earlier calculation, should be 25%.

Similarly, if you want a% of b, enter $a \times b$ and again press % rather than =. This time it is the division by 100 which is automatic. Try this one with the example 17.5% of 80 which we did above. Of course, the decision as to whether you need the × or the [÷] calculation is still up to you – you can't leave that to the calculator!

1.6.4 Arithmetic with percentages

(a) Increasing an amount by a fixed percentage

We have already mentioned this situation above, when we talked about adding 17.5% VAT to a bill. There is one general point, often missed by newcomers to this topic, which can save time and keystrokes in calculation: if you want to add a certain percentage to a quantity, there is no need to work out the increase and then add it on. For example, in the calculation above to add 17.5% VAT to a bill of £80, it is a waste of time to find 17.5% of £80 = £14, and then add this to the £80. Just notice that the original bill constitutes 100%, and we are now adding a further 17.5%, so altogether we will end up with 117.5% of the original amount. All we need do, then, is find 117.5% of £80 = £94 – a much quicker operation.

Similarly, if you want to add 10% mark-up to a purchase price, multiply the price by 110% = 110/100, and so on. We can generalise this to say:

- **To add x% to an amount a, we find $(100 + x)$% of a.**

(b) Reducing an amount by a fixed percentage

Sometimes, instead of adding on a percentage, we want to take it off. A simple example arises in fixing sale prices, where you will often see statements such as '25% off'. Suppose the original price of an item was £84, and we need to reduce it by 25%. One possible method, of course, would be to find 25% of £84 and then subtract it from £84; if you do this you should get the answer £63.

However, a better way is to note that the original price represents 100%, so that when we take off 25% we are left with 75%. Thus finding 75% of the original price will have the same effect as taking off 25% – and indeed, if you check this you will find that the result is £63 as before.

In general, then:

- **To *reduce* an amount by x%, we need to find $(100 - x)$% of the amount.**

(c) Accumulating percentage changes

One important area of application of percentages is in the calculation of interest. Interest rates on investments, borrowing and so on are almost always quoted in percentage terms. Thus we are accustomed to reading in our newspapers that 'the Chancellor of the Exchequer has cut interest rates by half a per cent'.

Sometimes, in order to make short-term borrowing rates look less alarmingly high, a company will quote a rate per month rather than per annum. In this situation it is important to understand how such percentage charges accumulate, and what will be the net effect over a year of a given interest rate per month. Many people, for example, borrowing a sum for three months and being told that a rate of 5% per month is charged, will mistakenly multiply this by 3 and assume that the overall rate they pay is going to be 15%. That sounds bad enough, but in fact, as we will discover, it is something of an underestimate.

To see why this is so, suppose the sum borrowed was £100. By the end of the first month you owe the original £100 plus 5% of £100 – that is 105% of £100, or £105 altogether. If you do not repay the sum, you will then incur interest for the second month on this whole amount – 105% of £105, (which is the same as 105% of 105% of £100), giving a debt of £110.25. Finally, in month three you incur further interest on this whole sum, so that your final debt is 105% of £110.25, which amounts to £115.76. Thus the true rate of interest for the three-month period is not 15%, but 15.76%. This may not seem like a big difference, but over a longer period, or with a larger sum borrowed, the discrepancy could be considerable.

The reason our original estimate of 15% for the interest rate over 3 months was wrong is that it assumed that each month we were only paying interest on the original £100 borrowed, whereas, of course, we actually pay interest in month 2 on the £5 interest added in month 1, as well as the £100 – and so on. This is known as *compounding*, and is the reason why we cannot simply multiply the rate for 1 month by 3 to get the rate for 3 months. We will spend some time in Chapter 6 looking at more problems related to interest, but for the time being simply note that you need to be careful when scaling percentages up or down over time.

1.6.5 Percentage to decimal conversions

Sometimes we need to convert percentages to decimals, or vice versa. The process is simple:

- **To express a percentage as a decimal, move the decimal point two places to the left (i.e. divide by 100).**
- **To express a decimal as a percentage, move the decimal point two places to the right (i.e. multiply by 100).**

1.6.6 Use and misuse of percentages

We need to be careful about the statements we make when comparing percentages. The Retail Prices Index (RPI), for example, expresses the cost of goods in the current month as a percentage of the cost in a fixed *base month*. But if the RPI increases from 256 to 267, we cannot say 'there has been an 11% increase', since this would usually be interpreted as meaning 11% of 256, which is, of course, untrue. All we can justifiably claim is that the RPI has risen by eleven *percentage points*.

Another common problem, which you can often spot in the popular press, is the addition or subtraction of percentages which do not relate to the same base figure. Consider, for example, the statement that 'so convenient is the Gatwick London Terminal that almost 25% of BA's premium passengers are checking in there; around 16% of long haul and 9% of short haul' (taken from BA's *Business Life* magazine). It seems likely that the 25% has been arrived at by adding the 16% and the 9%. But the 16% is based on the number of long-haul passengers, while the 9% is based on the (presumably much larger) number of short-haul passengers, and therefore the two figures may not be added, as we will demonstrate.

Suppose, to make things simple, that in a particular period of time there were 80 000 long-haul and 120 000 short-haul premium passengers. Using the percentages quoted, we conclude that of these passengers, 16% of the 80 000 – that is 12 800 – and 9% of the 120 000, which is 10 800, would use the London Terminal. In total, then, 23 600 out of 200 000 passengers use the terminal, and you can verify that this constitutes 11.8% – a very long way from the 25% given in the quote! (The 11.8% is, as many of you will realise, not the *sum*, but the *weighted average* of 16% and 9%:

$$\frac{80\,000}{200\,000} \times 16\% + \frac{120\,000}{200\,000} \times 9\%,$$

which is just another way of writing the calculation we did above.)

Yet another source of confusion arises from the mistaken belief that percentages can never be greater than 100. In some situations, of course, this is true; the statement in a recent issue of *Computing* magazine, that 20% of network adaptor cards (a technical bit of computing kit) cause 150% of problems is clearly nonsensical. But take the situation where a new university course attracted only 12 students in its first year of existence, but 30 in its second; then it is perfectly correct to say that the number of students has increased by 150%. The increase was 18 students, and this as a percentage of the first year's intake is 150%. Another way to convey the same information is to say that the second year's intake was 250% of the first year's – in other words, two and a half times as big.

We do not need to give further examples of this kind of misuse of percentages – though there is unfortunately no shortage of examples; you may enjoy spotting them for yourself. However, we hope we have said enough to alert you to the fact that percentages are not quite as simple as they look, and that, both as a user and a consumer of percentage-based data, you need to take care.

Exercises 1.5

1. What is 63 as a percentage of 150?
2. What is 30% of 120?
3. A construction company bills a customer for £22 450. However, a penalty clause in the contract makes the company liable to a reduction of 12% for late completion. How much will it actually receive?
4. What will a price of £2.50 become when a 16% mark-up is added?
5. Express 2% as a decimal.
6. Express 0.225 as a percentage.
7. A firm has 237 male and 161 female employees. What percentage of its workforce is male?
8. Last year 346 customers of a small building society defaulted on their mortgages; this year the figure was 279.

By what percentage has the number of defaulters decreased? Do you think this a useful measurement? If not suggest a more useful one, and indicate what further information you would need to have in order to compute it.

9. In 1953, when Queen Elizabeth II came to the throne in Britain, there were 1 087 000 people employed in agriculture, forestry and fishing; in 1993 the figure had fallen to 279 000. By what percentage did the number employed in this industrial sector fall over the 40-year period?

10. If there is a steady growth of 3% per annum in the population of a certain town, by what percentage will its population increase over a ten-year period? (You may find it easiest to examine the effect of this rate of growth on a hypothetical population of, say, 10 000 people.)

1.7 Some New Arithmetic Operations

We have been using the symbols for familiar arithmetic operations such as addition, multiplication and percentage more or less without comment. However, we now want to introduce some operations, and their symbols, which may be new to you, but which you are certain to come across if you are about to study a statistics or quantitative methods course. Although they may seem confusing to begin with, with practice they will soon become part of your everyday quantitative vocabulary – after all, a division sign probably seemed a pretty strange symbol when you first met it!

1.7.1 *The summation sign*

Many operations, like finding an average, involve the addition of series of figures. Because it is a nuisance to have to say 'add all these numbers together' in words, we introduce a special symbol Σ, meaning 'add together'. The symbol is a Greek capital sigma, or S, and stands for 'sum'. Thus if we have the takings of a series of branches of a retail chain, and want to find the total takings for all branches, we can express the required calculation as Σ(takings

Better still, if there are 12 branches and the takings of branch number 1 are written as t_1, those of branch 2 as t_2, etc, then we can write:

$$\sum_{i=1}^{12} t_i$$

showing that we want to add all the ts from t_1 up to t_{12}. When the range over which we are adding up is clear from the context, we sometimes leave out the subscripts and just write:

$$\sum t$$

pronounced 'sigma t'.

You will find the sigma sign used a great deal in statistical formulae; indeed we shall be using it in that context in Chapter 7 of this book.

7.2 The modulus sign

Sometimes we are interested only in the *size* of a figure, and not in whether it is positive or negative. This might be the case, for instance, if we are examining the accuracy with which we have managed to forecast the demand for stocks of a commodity at a warehouse. If we are simply interested in how far 'out' the forecasts have been, and not in whether they were high or low, then a forecast which was too low by, say, 50 (giving an error of +50) and one which was too high by 50 (with a resulting error of −50) would be treated in the same way.

We call the size of a number x without its sign the *absolute value* of x, and denote it by $|x|$. Thus $|-50| = |50| = 50$. If you want a general definition, we could say

$|x| = x$ if x is positive or zero.
$|x| = -x$ if x is negative.

But it is probably easiest simply to think of the rule as 'ignore the sign'. Another name for the absolute value is the *modulus*. You will find that many computer packages, such as spreadsheets, will calculate the modulus with a function such as ABS(X).

1.7.3 The factorial sign

Sometimes – for example, in certain probability calculations which you may encounter – we need to multiply together all the numbers less than a given figure: for example, we need to find $5 \times 4 \times 3 \times 2 \times 1$, or $3 \times 2 \times 1$. Such expressions are called *factorials*, so that $5 \times 4 \times 3 \times 2 \times 1$ would be referred to as 'five factorial', and so on. To abbreviate the writing of factorials, we use the symbol !, writing 5! instead of $5 \times 4 \times 3 \times 2 \times 1$. You may find that your calculator has a key for evaluating factorials.

The expression $n!$ can only be defined in this way when n is a positive integer, but one special case needs mentioning: we define 0! to be equal to 1.

1.7.4 Inequalities

We are all accustomed to using the sign for equality, [=] as used in the expression $3 \times 4 = 12$ to show that the quantities on the left and right of the sign are equal. However, in many practical situations we need a way of showing, not that two quantities are equal, but that one is bigger or smaller than another. For example, a course may have a limit of 50 students, so that the number of students recruited must be no more than 50; we may have advance orders for 120 of a certain product, so that number produced must be at least 120; and so on.

There are four signs used to represent these *inequalities*, as they are called:

- $a < b$ means 'a is less than b'.
- $a > b$ means 'a is greater than b'.
- $a \leq b$ means 'a is less than or equal to b'.
- $a \geq b$ means 'a is greater than or equal to b'.

In each case the larger end of the sign is directed towards the larger quantity. You are perhaps more likely to encounter the last two signs in practical applications. So we could express the statements in the last paragraph as 'number of students $\leq$ 50' and 'number of items produced $\geq$ 120'.

The $\leq$ sign is particularly useful when we want to specify that a number must lie within a certain range. The statement 'we need at least two people to work on this project, but no more than eight', for example, could be written as '$2 \leq$ number of people ≤ 8'.

Exercises 1.6

1. If x takes the values 1, 3, 5, 7, 9, what is the value of $\Sigma\, 3x$?
2. What is (a) $|-3|$; (b) $|8|$; (c) $|7-11|$?
3. Which of the following statements are true?
   (a) $2 < 3$ (b) $-1 > -7$ (c) $6 \leq 8$ (d) $3 - 5 \geq 0$?
4. If $-2 < y < 4$, and y is an integer, list all possible values of y.
5. Use the appropriate inequality sign to write an expression indicating that the selling price of an item must never be more than one-and-a-half times its cost.
6. Compute the following expressions:
   (a) $6!$; (b) $10!/5!$; (c) $4! \times 0!$

2 Algebraic Expressions

2.1 Introduction

Arithmetic is all we need if we are always operating on quantities whose numerical values are known. For example, in preparing the profit and loss statement of a company for a given financial year, arithmetic is all that is necessary. The values of sales, interest charges incurred, depreciation charges and so on are all known quantities for the year in question. All the accountant needs to do is to perform the right arithmetical operations on these quantities to arrive at the profit or loss figure for the year.

Very often, however, we do not know the numerical values of the quantities on which we wish to operate. This could be for a number of reasons. In the business context the most usual reason is that the quantity will only be known in the future, or it can take any one of a number of values and we do not know at the outset what value to give it to best advantage.

Take, for example, a potato crisp manufacturer. The price s/he will be charged per ton for next year's supply of potatoes is an unknown quantity. The decision as to how many tons of potatoes to contract to buy is likely to depend on the price the supplier demands. So the amount to be bought is also an unknown quantity. We can see now that arithmetic will fail to help us say anything very useful about the total cost of potatoes to the crisp manufacturer next year.

If we assume the crisp manufacturer contracts to buy, say, 10 tons of potatoes at £500 per ton, his or her costs will be £5000. Statements of this kind are as far as arithmetic will enable us to go. But such statements do not help us to say in a compact and yet general way what the cost of potatoes to the crisp manufacturer will be. He or she might buy any number of tons, at any price.

Algebra helps us to say in a precise way what the price paid by the manufacturer for potatoes will be next year. It does this by letting *variables* stand for unknown quantities. Thus, if £P is the price per ton the supplier charges and T is the quantity in tons that the manufacturer contracts to buy, then his or her cost will be £C, where

$$C = P \times T. \tag{2.1}$$

Of course, we still do not know the total cost C of potatoes. This we will not know until we know the numerical values for both P and T. However, the advantage in writing (2.1) is that it enables us to have a compact and precise representation of the cost for any price P and any quantity T. We can now use C as written in (2.1) in any further analysis we may wish to undertake of the situation. For example, if the supplier asks for 10% of the contract value to be paid upon signature of the contract then we can say that the crisp manufacturer must pay £$0.1 \times C$ or £$0.1 \times P \times T$ on signing the contract. Thus we now have a compact representation of the amount payable upon signature. Such compact representations and manipulations are impossible just by using arithmetical operations, unless T and P are known.

We have been able to generalise the cost calculation for the crisp manufacturer by using *variables* to represent the components of the cost. This is the key to how *algebra* extends the power of arithmetic. Variables in algebra are grouped into *algebraic expressions*.

● **Algebraic expressions consist of numbers and variables linked by symbols indicating arithmetical operations.**

The product $P \times T$ in (2.1) is an example of an algebraic expression. It says that the values of P and T are to be multiplied together. Another example is $0.1 \times C$, referred to earlier. Algebraic expressions are the vehicle we use to translate verbal descriptions of relationships between quantities into mathematical terms. Once this has been done the situations can be analysed mathematically by manipulating the algebraic expression(s).

One more simple example will help demonstrate how necessary algebra is when unknown quantities are involved in decision situations.

Essential Mathematics

EXAMPLE

A firm distributed £2m in dividends last year and wants to raise this amount by up to 10% in the current financial year. The firm is liable to pay tax at 20% of the amount distributed as dividends. How much will the firm's tax liability be in respect of dividends distributed during the current year?

SOLUTION

Arithmetical approach

If the firm does not raise the amount distributed as dividends, then since 20% = 0.2,

tax liability = £2m × 0.2 = £0.4m = £400 000.

If it raises the amount distributed by the full 10%, dividends distributed will be £2 × 1.1 = £2.2m and

tax liability = £2.2m × 0.2 = £0.44m = £440 000.

Clearly, the firm's tax liability will be somewhere between £400 000 and £440 000 depending on how much it decides to increase the dividends.

How can we link the increase in dividends to the consequent tax liability? Using calculations similar to those shown above for a 10% rise in dividends, we can compute the tax liability for a 1%, 2%, etc. rise in dividends distributed. The results are shown in Table 2.1.

TABLE 2.1
Increase in dividends linked to tax liability

Increase on last year's dividends	Tax liability
(%)	(£000)
0	400
1	404
2	408
↓	↓
10	440

Clearly, this is a time-consuming and cumbersome way to compute the tax liability. More to the point, however, it is not precise or complete. For example, what if the firm decided on a 1.5% rise in dividends over last year? The tax liability for this rise is not tabulated. Similarly the table does not enable the firm to explore its tax liability for raising dividends by more than the 10% initially contemplated.

Many of you will, of course, have spotted that every percentage point rise in dividends increases tax liability by £4000. This provides the means to compute the tax liability for any rise in dividends. For example, if dividends rise by 5.5% on last year's £2m, then last year's tax liability of £400 000 will rise by 5.5 × £4000 = £22 000, making the total tax liability this year £422 000.

However, it is not necessary to construct a table as above to arrive at a way of computing the tax liability for any given increase in dividends. Indeed, in more complex cases the table may give no clue to the formula underlying the values within the table. Algebra helps us to get *directly* at the underlying formula without the intermediate step of the table.

Algebraic approach

The key to this approach, as in the earlier crisp-manufacturing example, is to use variables for unknown quantities. In the case of the firm in the example we have two unknowns: the rise on last year's dividends and the consequent tax liability. Thus let

> INCREASE be the percentage rise on last year's dividends;
> and
> TAX be the consequent tax liability in (£000).

Last year's dividends were £2m, or 2 000 000 pounds. If this amount is increased this year by INCREASE%, dividends this year, in thousands of pounds, will be:

$$£2000 + \frac{\text{INCREASE}}{100} \times £2000.$$

Since tax is levied at 20% on the amount distributed as dividends the corresponding tax liability will be:

$$\text{TAX} = 0.2 \times £2000 + 0.2 \times \frac{\text{INCREASE}}{100} \times £2000.$$

We can simplify this expression to:

$$\text{TAX} = £400 + 4 \times \text{INCREASE}. \tag{2.2}$$

The formula in (2.2) is a much more general statement of the tax liability than we could make by relying on arithmetic alone. Remembering that it shows the tax liability in thousands of pounds we can see that if the firm keeps dividends at last year's levels (i.e. if INCREASE = 0) the tax liability is £400 000. For every percentage point rise of dividends from last year's levels (i.e. for every unit rise in the value of INCREASE) the tax liability rises by 4 thousand pounds. This enables us to compute the tax liability for any increase (or indeed reduction) in dividends the firm might decide. (Remember that a negative increase is a reduction in dividends.)

Algebraic expressions are the fundamental building blocks of algebra. In this chapter we look at the rules which govern their construction and manipulation; this lays the groundwork for applications of algebra to practical problems later in the book.

2.2 Powers and Roots

2.2.1 Introduction

Powers and roots are often used within algebraic expressions. We shall see in this section what they stand for and how we can carry out operations with them.

Let us take powers first. An example will help illustrate what a power is.

EXAMPLE

A regional branch of an insurance company employs 10 salespersons. Each salesperson is set a target to sell 10 pensions each month. If all salepersons meet their target exactly how many pensions will the branch sell in a period of 10 months?

SOLUTION

Clearly, the answer is:

10 (salespersons) × 10 (pensions each) × 10 (months)
= 1000 pensions.

We have multiplied 10 three times by itself to find the total pensions sold. This we would normally write as:

pensions sold = 10 × 10 × 10.

Powers enable us to write this sort of repeated operation in a shorthand way as pensions sold = 10^3. The expression 10^3 is called a *power* of 10. It says that 10 is to be multiplied by itself three times. This is called the third power or *cube* of 10.

We can generalise this notation from numbers to variables. The meaning is exactly the same. For example we may write a^2 to mean $a \times a$; this is usually read as 'a squared'. Similarly a^5 means $a \times a \times a \times a \times a$, and is read as 'a to the power 5'.

The notation can also be used with algebraic expressions such as $(2a)^4$. This would mean the expression $2a$ is to be multiplied by itself four times. So:

$$(2a)^4 = 2a \times 2a \times 2a \times 2a.$$

Generally, when n is a positive integer the power a^n means a multiplied by itself n times; that is,

$$a^n = a \times a \times a \ldots a, \qquad (2.3)$$

a appearing n times after the = sign.

In talking about a^n we call a the *base* and n the *exponent* or *index* of the power. The power itself is read as 'the nth power of a', or 'a to the power n'. The process of multiplying a by itself n times is called 'raising a to the power n'. Sometimes the index on its own is loosely referred to as the power.

Once we know what powers are we can easily define *roots*. Recall from Chapter 1 the idea of an inverse operation (for example, division was the inverse of multiplication). A root is shorthand for the inverse of the operation of raising to a power. The symbol for a root is $\sqrt{}$. For example, $^2\sqrt{9}$ stands for the operation of finding the number which gives the result 9 when it is raised to the power 2. The number is 3, and so we say *the second or square root of 9 is 3*. Similarly, $^3\sqrt{8}$ stands for the number which gives 8 when it is raised to the power 3. The number in this case is 2. So we say the third or cube root of 8 is 2. If no number is outside the $\sqrt{}$, $^2\sqrt{}$ is assumed. Thus $\sqrt{25} = 5$.

We can generalise the foregoing by saying that

- **The nth root of a quantity, written $^n\sqrt{a}$, is a quantity b such that $b^n = a$.**

EXAMPLES

Find the following roots:

$^3\sqrt{64}$

$^2\sqrt{a^2}$

SOLUTION

The cube root of 64 is the number which would give 64 when raised to the power 3. That number is 4. Thus $^3\sqrt{64} = 4$.

The square root of a^2 is the expression which when raised to the power 2 gives a^2. That is, of course, a and so we have:

$$^2\sqrt{a^2} = \sqrt{a^2} = a.$$

There are two points relating to even roots which we need to clarify before we leave the definition of roots (by *even roots* we mean roots such as the second, fourth and so on).

First, when we take an even root we do not get a unique result. For example, one square root of 64 is 8. However, another one is -8 because $(-8) \times (-8) = 64$. We generally write the two square roots of a number using the sign $\pm$, so that for example $\sqrt{64} = \pm 8$, $\sqrt{81} = \pm 9$, and so on. Second, a negative number cannot have an even root. For example $^2\sqrt{-4}$ is not defined because there is no number we can square to get the result -4. (There is a system of 'imaginary' numbers used by scientists where a negative number can have an even root. This system of numbers, as distinct from the system of 'real' numbers we are all familiar with, is beyond the scope of this book.)

We have defined powers and roots so far with reference to positive integer exponents only. We will see how we can interpret exponents which are fractional and/or negative after we have seen how we can operate with powers.

Exercises 2.1

1. Evaluate the following expressions:
   (a) 14^2; (b) $\sqrt{625}$; (c) 6^4;
   (d) $\sqrt[3]{125}$; (e) $\sqrt[3]{(-343)}$; (f) $\sqrt[4]{(-256)}$.
2. Simplify the following roots:
   (a) $\sqrt[3]{(a+8)^3}$; (b) $\sqrt[2]{(a-2)^2}$; (c) $\sqrt[6]{(-2x)^6}$.

2.2.2 Operations with powers

Powers can be manipulated with the aid of a few simple rules.

(a) Addition and subtraction of powers

The simplest power we can have is the first, as in a^1, normally written as just a. We may now ask what meaning can be attached to the sum of two such powers, $a + a$. You can see what the answer must be by considering an arithmetical example such as $3 + 3 = 6 = 2 \times 3$: adding 3 to itself is equivalent to multiplying 3 by 2. In the same way $3 + 3 + 3 = 3 \times 3$, and so on – this is, of course, how we define multiplication. So in an analogous way, we write $a + a = 2a$, $a + a + a = 3a$, etc.

We can then extend this idea to say that $a + 2a = 3a$. The 2 and 3 here are called *coefficients* of a; there is also an unwritten coefficient of 1 in front of the first a in this expression. So in performing this addition, we are simply adding the coefficients of a. Similarly $6z - 4z = 2z$. However, we cannot simplify $6a - 4z$, since the two terms involve two different variables.

There is, in fact, nothing special about a^1; in exactly the same way we can say $a^2 + 3a^2 = 4a^2$. The terms a^2 and $3a^2$ are called *like* powers, the power being a^2. The coefficients here are the numbers 1 and 3, but in another case they might themselves be algebraic expressions. We cannot simplify $a^2 + a^3$ because a^2 and a^3 are not *like* powers.

We can sum up this discussion by saying:

- **In order to add or subtract like powers we simply add or subtract their coefficients.**

For example,

$$3a^2 + b^3 - a^2 + 4b^3 = 3a^2 - a^2 + b^3 + 4b^3$$
$$= (3 - 1)a^2 + (1 + 4)b^3 = 2a^2 + 5b^3.$$

(Notice how we use brackets to clarify that in $(3 - 1)a^2$, we first subtract 1 from 3, and then multiply by a^2. This is an extension of the use of brackets which you met in Chapter 1.)

Had the like powers had algebraic expressions for coefficients the above rule would have been applied just the same. For example,

$$3a^4 + b\,a^4 = (3 + b)a^4.$$

The like power is a^4, so when its coefficients are added the resulting coefficient becomes $3 + b$.

We have been making a tacit assumption in this discussion that, in interpreting an expression such as $3a^2$, the a is squared *before* it is multiplied by the 3. In other words, in terms of the priorities of operations,

- **Exponentiation (that is, raising to a power) takes precedence over multiplication and division (and therefore, of course, over addition and subtraction).**

(b) Multiplication of powers

If the powers we wish to multiply have different bases then there is little we can do by way of simplification unless we can actually evaluate some of the powers. For example, we cannot further simplify the product $a^2\,b^3$, but $2^2\,b^3$ can be simplified to $4b^3$.

If, however, the powers we wish to multiply have the same base then we *can* simplify the product. For example if we want to multiply a^3 by a^2 we can write the product as a^5. To see this, note that:

$$a^2 \times a^3 = (a \times a) \times (a \times a \times a) = a^5 = a^{2+3}.$$

We can generalise this to give a rule for multiplying powers of the same base:

- **To multiply powers, add their indices.**

This rule can be expressed more succinctly as:

$$a^n \times a^m = a^{n+m}.$$

Some examples to illustrate this rule are the following:

$$5 \times 2^4 \times 2^4 = 5 \times 2^8.$$
$$a^3 \times ab^4 = a^4 b^4.$$
$$2^3 \times a^2 \times 2 \times 2^2 \times a^4 = 2^6 \times a^6 = 64a^6.$$

(c) *Division of powers*

As with products of powers, a quotient of powers which have different bases cannot be simplified unless we can evaluate some of the powers.

Where, however, we wish to divide powers which have the same base then we can simplify the division. For example, if we divide 6^5 by 6^2 we will find the result is 6^3. We can deduce this by using the definition of a power. Thus,

$$6^5 \div 6^2 = (6 \times 6 \times 6 \times 6 \times 6) \div (6 \times 6) = (6 \times 6 \times 6)$$
$$= 6^3 = 6^{5-2}.$$

This observation can be generalised to give the following rule for dividing powers which have the same base:

- **To divide powers, subtract the index of the denominator from that of the numerator.**

We can say much more simply,

$$a^n \div a^m = a^{n-m}.$$

Some examples of this rule are the following:

$$6 \times 2^4 \div 2^2 = 6 \times 2^2 = 24.$$
$$ab^4 \div b^3 = ab$$
$$(2^3 \times a^2 \times 2) \div (2^2 \times a) = 2^2 \times a = 4a.$$

2.2.3 Extensions of the rules for operating with powers

The rules we have introduced are sufficient for dealing with any operations on powers you might have to perform. Certain extensions of the rules, however, come in very handy for operating on more complex powers. These extensions are summarised below.

(a) Raising a product to an exponent

Suppose we wanted to raise the product of two factors, say $3a$, to the power 2. The expression for this operation is, of course, $(3a)^2$ (we need the brackets to show that both the 3 and the a are to be squared). Can we simplify this at all? Resorting to the rule for multiplying powers of the same base we can write

$$(3a)^2 = (3a)(3a) = 3a \times 3a = 9\,a^2 = 3^2 a^2.$$

So squaring $3a$ results in a product of 3^2 and a^2.

This observation can be generalised:

- **To raise a product to an exponent simply raise the individual factors to the exponent.**

This rule can be summarised in mathematical terms as

$$(abc\ldots)^n = a^n\, b^n\, c^n \ldots$$

For example,

$$(4bc)^2 = 4^2\, b^2\, c^2 = 16b^2\, c^2$$

or

$$(3 \times 4)^3 = 3^3 \, 4^3 = 27 \times 64 = 1728 \ (= 12^3, \text{ of course}).$$

It is worth stressing that the above rule only holds for products and *not* for sums or differences being raised to an exponent. For example it is not correct to expand $(a + b)^2$ into $a^2 + b^2$. You can verify this by noting that $(3 + 1)^2 = 4^2 = 16$, whereas $3^2 + 1^2 = 9 + 1 = 10$. If you are tempted to perform algebraic operations about which you are uncertain, it is often a good idea to try them out with simple numbers like this, and see whether the results are correct.

(b) Raising a power to an exponent

If we need to raise a power, say a^2, to an exponent such as 3, the expression for the operation to be performed would be $(a^2)^3$. We can simplify this expression just as we did above by resorting to the rule for multiplying powers. We have

$$(a^2)^3 = a^2 \times a^2 \times a^2 = a^6 = a^{2 \times 3}.$$

Thus raising a^2 to the exponent 3 is equivalent to simply raising the base a to the power 2×3. This observation can be generalised to give the rule:

$$(a^m)^n = a^{mn}.$$

For example,

$$(c^3)^2 = c^6;$$
$$(5^{1/3})^9 = 5^{(1/3)9} = 5^3.$$

(c) Raising a fraction to an exponent

In order to raise a fraction such as 6/2 to a power, remember that you can raise the numerator and the denominator to the power first and then carry out the division. Thus,

$$(6/2)^2 = 6^2/2^2 = 36/4 = 9.$$

You can verify this is the correct result by noting that:

$$(6/2)^2 = 3^2 = 9.$$

So, in general,

- **To raise a fraction to an exponent we raise both the numerator and the denominator to the exponent.**

Or, more simply,

$$(a/b)^n = a^n/b^n.$$

So if we want to raise the fraction 4/5 to some exponent, say 2, we would have

$$(4/5)^2 = 4^2/5^2.$$

Similarly,

$$(15/4)^2 = 15^2/4^2 = 225/16;$$
$$(ab/c)^5 = (ab)^5/c^5 = a^5\,b^5/c^5.$$

2.2.4 *Powers where the exponent is not a positive integer*

We can use the operations with powers we have introduced to interpret powers where the exponent is not an integer.

(a) Fractional exponents

Take, for example, $9^{1/2}$. We can write 9 as 3^2 and so we have:

$$9^{1/2} = (3^2)^{1/2}.$$

Using the rule for raising a power to an exponent we can deduce that

$$(3^2)^{1/2} = 3^{2 \times 1/2} = 3.$$

Hence, we deduce that

$$9^{1/2} = 3$$

and so $9^{1/2}$ and $\sqrt{9}$ are the same thing, namely 3, the square root of 9.

This equivalence between a root and a fractional power is by no means unique to square roots. For example:

$$8^{1/3} = (2^3)^{1/3} = 2^{3 \times 1/3} = 2$$

and so $8^{1/3}$ is equal to the cube root of 8, so that $8^{1/3} = \sqrt[3]{8} = 2$.

We can extend our discussion to cover fractional exponents with numerators other than 1.

EXAMPLE

Evaluate $4^{3.5}$.

SOLUTION

We can write $4^{3.5}$ as $4^{7/2}$.
Hence $4^{3.5} = 4^{7/2} = (4^7)^{1/2}$.
This means $4^{3.5}$ is in fact the square root of 4 after it is raised to the power 7.
Hence $4^{3.5} = \sqrt{4^7} = \sqrt{16384} = \pm 128$.

We can summarise this discussion by saying that when n and m are positive integers,

$a^{m/n}$ means $\sqrt[n]{a^m}$.

From this interpretation of a power with a fractional exponent you should be able to deduce that powers and roots are the *same* thing:

A root is a power with a fractional exponent.

For example,

$$\sqrt{49} = 49^{1/2} = 7^{2 \times 1/2} = 7.$$

$$12^{1.5} = 12^{3/2} = \sqrt{12^3} = \sqrt{1728} = \pm 41.57.$$

b)   Powers where the exponent is zero

The rule for dividing powers can lead to a power with an exponent of zero. For example, the division of 3^2 by 3^2 gives $3^{2-2} = 3^0$. We know, of course, that dividing 3^2 by 3^2 is the same as dividing 9 by 9 and the quotient is 1. This leads us to the conclusion that $3^0 = 1$. We could have chosen any power to divide by itself and the result would still have been 1. Hence:

- **Any power with an exponent of 0 equals 1.**

In other words,

$$a^0 = 1$$

where a is any number or algebraic expression.

For example,

$$140^0 = 1;$$
$$[ab - (cb - 3c^3)]^0 = 1;$$
$$(ab)^0 + [ab - (bc - 1)^2]^0 = 2, \text{ etc.}$$

(c) Powers where the exponent is negative

The observation above that anything to the power zero equals 1 can help us interpret a power with a negative exponent. Take, for example, 5^{-2}. We can show that this equals $1/5^2$ or simply $1/25$.

To see this, note that we can write $1/25 = 5^0/5^2$ as we know that $5^0 = 1$. But by the rules for dividing powers we also know that $5^0/5^2 = 5^{0-2} = 5^{-2}$. So we conclude that:

$$5^{-2} = 5^0/5^2 = 1/5^2.$$

So we can interpret the power with a negative exponent as a fraction with 1 in the numerator and the power with positive exponent in the denominator.

It is much simpler to express this rule in mathematical notation as:

$$a^{-n} = 1/a^n.$$

For example,

$$2^{-2} = 1/2^2 = 1/4;$$
$$4^{-1/2} = 1/4^{1/2} = 1/2 \text{ or } -1/2;$$
$$(a - b)^{-2/3} = 1/(a - b)^{2/3} = 1/\sqrt[3]{(a - b)^2}.$$

2.2.5 Potential pitfalls to be avoided

It is worth highlighting some common sources of error in handling operations with powers:

(i) We saw earlier that when we raise a sum to a power, as in $(a + b)^2$, it is not correct to say $(a + b)^2 = a^2 + b^2$. However, it *is* correct, as we saw, to say for a product that $(ab)^2 = a^2 b^2$.
(ii) It is not right to say that $\sqrt{(a^2 + b^2)} = \sqrt{a^2} + \sqrt{b^2}$ but it is, of course, true to say $\sqrt{(a^2 b^2)} = \sqrt{a^2} \sqrt{b^2} = ab$, since $\sqrt{(a^2 b^2)} = (a^2 b^2)^{1/2} = (a^2)^{1/2} (b^2)^{1/2} = ab$.
(iii) It is correct to say that $a^{-2} + b^{-2} = 1/a^2 + 1/b^2$, but **incorrect** to say that $a^{-2} + b^{-2} = 1/(a^2 + b^2)$.

Exercises 2.2

1. Evaluate the following:
   (a) $(-25)^3$;   (b) $\sqrt{625}$;   (c) 10^0;
   (d) 5^{-1};   (e) $\sqrt{5^3}$;   (f) $(5/2)^2$.
2. Evaluate the following expressions:
   (a) $(2/5)^2$;   (b) $\sqrt[3]{5^6}$;   (c) 16×2^{-2};
   (d) $(25^{1/2}\, 8^{1/3})^{-1}$;   (e) $(8^{-1/3} + (2/7)^{-1})^{1/2}$;
   (f) $3(-5)^4$;   (g) $\sqrt{(-2)^{-6}}$.
3. Simplify the following expressions by removing negative exponents:
   (a) $(cd^{-4} + a/b^2)^{1/2}$;   (b) $\sqrt[4]{(1 + b)^{-4}}$;
   (c) $(ca)^2\, c^{-2}\, b^2\, a^{-2}$;   (d) $(2a^2 - 2a^{-1}\, a^3)^2$;
   (e) $(2a^{-2}\, 2^{-1} + 1)^2\, (1/a + 1)^{-2}$.

2.3 Using Brackets in Algebra

2.3.1 Introduction

We have used brackets so far to contain algebraic expressions to be raised to a power, or to separate operating symbols in an expression such as $3 \times (-2)$. In this section we will discuss in more detail why brackets are used in algebra and how algebraic operations are affected by them.

A simple rule on the use of brackets is sufficient to show why they are so useful. This rule is:

- **Operations within brackets are performed before operations outside.**

This means we can use brackets to change the conventional priority of operations. This change becomes necessary in situations such as the following example.

EXAMPLE

An accounting firm must audit the annual accounts of six of its clients next month. One client is a large company requiring 200 hours of an auditor's time. The remaining five clients are of similar size and their accounts will average 30 auditor hours each. The firm has 10 auditors. If work on the six clients is to be shared equally among the firm's auditors how many hours should each auditor be allocated?

SOLUTION

(i) The work on the accounts of the five small clients will take $5 \times 30 = 150$ hours.
(ii) Adding to this the work for the large client gives a total of $150 + 200 = 350$.
(iii) Sharing this work among the 10 auditors we have $350 \div 10 = 35$ hours each.

If we write the operations (i) – (iii) in a single algebraic expression without using brackets we would get:

$5 \times 30 + 200 \div 10.$

Following the conventional order of operations we would conclude from this expression that each auditor should be allocated 170 hours! As you can see, we would have failed to share the 150 hours from the small clients among the 10 auditors because 200 would have been divided by 10 *before* 150 was added to it. This was the correct thing to do, given the way we had written the expression, since division, by convention, has precedence over addition. Using brackets we can change this precedence to match what we actually know to be correct. To see this, write the operations (i) − (iii) above as:

$$(30 \times 5 + 200) \div 10.$$

The rule with brackets is that operations within them are performed first. So in the above expression we need to evaluate the interior of the bracket first. This gives 350. Then we can divide by 10 to end up with the correct answer of 35.

In summary, brackets are used in algebra to dictate a certain order of operations. They become necessary when the conventional order of operations would not lead to the right result. If you are uncertain whether the conventional order of operations without brackets leads to the right result, use brackets to clarify the order to be followed; it never does any harm to have too many brackets, even if they are not strictly required. It may prove necessary to have several sets of brackets nested within each other to prescribe the correct order of operations.

2.3.2 Removing brackets

The following rules need to be followed in removing brackets:

1. Perform operations within brackets first.

For example,

$$5 \times (10 - 3) = 5 \times 7 = 35.$$

This approach is not, however, possible if the expression within the brackets cannot be reduced to a single term. What proves useful in such a case is the following observation:

$$5 \times (10 - 3) = 50 - 15 = 35,$$

which is the same result as before.

This observation leads to the following general rule:

- **2. To remove brackets, multiply each term within the brackets by the factor outside the brackets**.

For example,

$$5(a - b) = 5a - 5b$$

or

$$-a(a - b) = -a^2 + ab.$$

One special case arises when we want to remove a set of brackets that appears to have no factor outside it.

For example, remove the brackets in:

$$5 + (10 - b).$$

Although there is no explicit factor written in front of the brackets here, there is nothing to stop us rewriting the expression as

$$5 + 1(10 - b).$$

We can now apply our rule, removing the brackets after multiplying each term inside them by +1. Thus we have:

$$5 + 1(10 - b) = 5 + 10 - b = 15 - b.$$

Clearly, in this special case the brackets can be simply discarded they were unnecessary in the first place.

A slightly more involved procedure is necessary to remove brackets when they are preceded by a minus sign. For example remove the brackets in:

$$10 - (4 - a + b).$$

We can think of this as being:

$$10 - 1(4 - a + b).$$

Thus we need to multiply the terms inside by -1 to remove the brackets, obtaining

$$10 - (4 - a + b) = 10 - 4 + a - b = 6 + a - b.$$

Note that if the brackets are preceded by a minus sign, in effect we simply change the signs of the terms inside as we remove the brackets.

One frequent case is where the factor outside the brackets is itself a bracketed expression. Rule 2 above is applied just the same: the factor multiplies every term inside the brackets. For example,

$$(ab + c)(c - d) = (ab + c) c + (ab + c) (-d).$$

As we still have brackets, we need to apply rule 2 again to remove them. We have

$$(ab + c) c + (ab + c) (-d) = abc + c^2 - abd - cd.$$

Another example is the following:

$$(a - 5)(a - b + 2c) = a(a - 5) - b(a - 5) + 2c(a - 5)$$
$$= a^2 - 5a - ab + 5b + 2ac - 10c.$$

So far, our rules have not dealt with nested brackets. Nested brackets are progressively removed by following the rule

- **3. Operations in the innermost brackets are carried out first.**

The following examples illustrate the use of this rule.

EXAMPLE A

Remove the brackets from $4(ac - b(b - c))$.

SOLUTION

The innermost bracket is $(b - c)$. To remove it, each one of its terms must be multiplied by the factor outside the bracket.

That factor is $-b$, so we have $-b(b - c) = -b^2 + bc$. Using this in the original expression gives $4(ac - b^2 + bc)$. Multiplying each term inside the brackets by 4 we remove them to obtain $4ac - 4b^2 + 4bc$.

EXAMPLE B

Remove the brackets from $b(ab - c(4 - c(d^2 - 2b)))$.

SOLUTION

Following the process in Example A and removing brackets from the innermost to the outermost we have:

$$b(ab - c(4 - c(d^2 - 2b))) = b(ab - c(4 - cd^2 + 2bc))$$
$$= b(ab - 4c + c^2d^2 - 2bc^2) = ab^2 - 4bc + bc^2d^2 - 2b^2c^2.$$

It's most unlikely, however, that you would have to deal with anything as complicated as this.

Exercises 2.3

1. Remove the brackets in the following expressions:
   (a) $2(10 - 4)$; (b) $-(6 - 8)$; (c) $a - 2(4 - b)$;
   (d) $4(b - a)$; (e) $-a(2a - b)$; (f) $(a - b)(a + b)$;
   (g) $b(c - b(cd + b))$; (h) $(a - b)[(2 - (b + a)3) - a(b - c)]$;
   (i) $(ab - cb)(4 - 2(bc - cd))$.
2. Use brackets in answering the following questions:
   (a) A carpenter always keeps some stock of timber. His orders to the supplier are always for 30 m³ of timber. He uses 15 m³ of timber a week. Use *I* for the inventory of timber in m³ the carpenter has and *N* for the number of orders the supplier has not yet delivered. Write an algebraic expression whose evaluation would lead to the number of weeks' work his inventory of timber plus that on order will cover.
   (b) An ice-cream manufacturer buys her strawberries from a farmer. She is offered them at £*P* a ton. However, if she buys more than 3 tons in the season she will pay

15% less per ton for any quantity in excess of 3 tons bought. Write an expression or expressions whose evaluation would give the cost of strawberries to the ice-cream manufacturer next year if she buys T tons in total.

(c) The normal return fare on an air route is £F. An airline operating the route offers a travel agent seats on each flight at a 10% discount for up to 20 seats. For every seat over 20 that the agent books the discount rises by 0.2% of F. The airline offers a maximum discount of 30% on £F. The discount applies to all seats booked by the agent on a flight and not just those that triggered a particular discount rate. If B is the number of seats booked by the agent on a flight, write one or more expressions whose evaluation would give the cost of the seats to the agent.

2.4 Factorising

Each term in a product is called a *factor*; for example, in $x^2 y$ the factors are x^2 and y, while in $(a + 1)(ab - 2)$, the factors are $(a + 1)$ and $(ab - 2)$. Factorising is the process of converting an algebraic expression into a product of two or more factors.

We will see later in the book how factorising comes in useful for solving equations. In this section we look at two simple methods of factorising: taking common factors and using identities.

2.4.1 *Factorising by taking common factors*

This is the simplest method of factorising. It can be seen as the reverse process to the one we followed for removing brackets with a factor outside.

For example, to remove the brackets from $c(ab + c)$ we multiplied both terms inside the bracket by c. This gave $abc + c^2$. Notice that each of the two terms of this expression is a product involving c; c is therefore said to be a *common factor* of the two terms.

We can reverse this process by using brackets. We divide each term by c, and place the resulting expression inside the brackets; c, the common factor, is placed outside the brackets. Thus

$$abc + c^2 = c(ab + c).$$

In this way the sum $abc + c^2$ has been factorised – that is, expressed as a product.

We can generalise this process as follows:

- **To factorise a sum or difference of terms which has a common factor, place the common factor outside brackets, and put inside the brackets the terms which remain after each has been divided by the common factor.**

For example,

$$a^2 + 2ab^2 = a(a + 2b^2);$$
$$ab^2 + 3cb^3 + ab^2c + 3b^3c^2 = b^2(a + 3cb + ac + 3bc^2).$$

Clearly, the key to factorising in this way is to find a suitable common factor. This may not always exist, so an algebraic expression may not be capable of being factorised. An expression may also be capable of being factorised in a number of different ways. For example, the second of the two expressions above can also be factorised as follows:

$$ab^2 + 3cb^3 + ab^2c + 3b^3c^2$$
$$= b^2(a + 3bc) + b^2c(a + 3bc)$$
$$= (a + 3bc)(b^2 + b^2c)$$
$$= b^2(1 + c)(a + 3bc).$$

So the expression has been written as three different products. Factorising will usually be done for a purpose, such as to simplify a quotient by cancelling. In such cases we know which one of the alternative products would be most useful.

2.4.2 Factorising by using identities

If you did question 1(f) from Exercises 2.3 you will have noticed that the product $(a + b)(a - b)$ when multiplied out gave the result $a^2 - b^2$. This is in fact the well-known identity for the *difference of two squares*, which, stated formally, is:

$$a^2 - b^2 = (a + b)(a - b).$$

It comes in very useful in a number of situations, as in the following examples.

EXAMPLE A

Factorise $x^2 - 16$.

SOLUTION

Noting that $16 = 4^2$, we can rewrite the expression as $x^2 - 4^2$. If you compare this with the general result above, you can see that it is a difference of squares with $a = x$ and $b = 4$. So using the formula gives:

$$x^2 - 16 = x^2 - 4^2 = (x + 4)(x - 4).$$

EXAMPLE B

Factorise $16a^4 - (a + b)^2$.

SOLUTION

$$16a^4 - (a + b)^2 = 2^4 a^4 - (a + b)^2$$
$$= (2^2 a^2 + a + b)(2^2 a^2 - a - b).$$

Two other useful identities for factorising are the following:

$$(a^3 + b^3) = (a + b)(a^2 - ab + b^2); \text{ and}$$
$$(a^3 - b^3) = (a - b)(a^2 + ab + b^2).$$

The following examples show the use of these identities:

EXAMPLE A

Factorise $a^4 - 8ab^3$

SOLUTION

$$a^4 - 8ab^3 = a(a^3 - 2^3 b^3)$$

Using the first of the two identities above to factorise the second bracket we have:

$$a^3 - 2^3b^3 = (a - 2b)(a^2 + 2ab + 4b^2)$$

so that altogether:

$$a^4 - 8ab^3 = a(a - 2b)(a^2 + 2ab + 4b^2).$$

EXAMPLE B

Factorise $c^3 + (2 + a)^3$

SOLUTION

$$c^3 + (2 + a)^3 = (c + 2 + a)(c^2 - c(2 + a) + (2 + a)^2).$$

You can always verify your factorising by multiplying out the product.

Exercises 2.4

Factorise the following expressions:
(a) $2a + 16b$;
(b) $3a^2 - 27ab + 21a^2b^2$;
(c) $2a - a^2 + 2b - ab$;
(d) $a^3 + a^2b - b^2a - b^3$;
(e) $27c^3 - 125$;
(f) $a^4 - 16(b + 2)^4$;
(g) $a^3 + 1$.

3 Functions and Graphs

3.1 Introduction

A *function* expresses a relationship between two or more quantities of interest. For example, we would expect the quantity of bread that people in a certain area will purchase to depend on the price of bread and the population of the area. To describe such a relationship we say that demand for bread is a *function* of the price and the population. Using D for the quantity demanded, P for the price of bread and S for the size of the population, we can write this function as:

$$D = f(P, S).$$

Writing the function in this way indicates that we believe the values of the variables D, P and S are related to one another, but we do not know the precise form of the relationship. Here D is known as the *dependent* and P and S as the *independent* variables – in other words, D takes values which are dependent on those of P and S.

In many cases we *do* know the exact form which the function takes. Generally this will enable us to write the function as an *equation* – that is, an algebraic expression involving the equals [=] sign. For example, we would expect the daily costs incurred by a car plant, C, to be a function of the number of cars made per day, N. Suppose the fixed daily costs are £1m and variable costs are £6000 for each car produced. Then C (£000) as a function of N is given by the equation:

$$C = 6N + 1000. \qquad (3.1)$$

In this case we have used equation (3.1) to express the fact that the quantities C and $6N + 1000$ must be equal.

There is one more important feature of a function that we must mention: for any value of the independent variable(s), we must be able to determine a *unique* value of the dependent variable. This is certainly the case with the function expressed by equation (3.1): for any given value of N there will be a single corresponding value of C. It is possible to have an equation in which not all value(s) of the independent variable(s) lead to a corresponding unique value of the dependent variable. Such an equation would not be a function. For example $y = \sqrt{x}$ gives not one but two values of y for every value of x: when $x = 9$, $y = 3$ or $y = -3$, and so on. This relationship is therefore *not* a function.

To sum up what we have said so far:

- **A function defines how one variable, known as the *dependent* variable, is uniquely determined by the values of one or more other variables, known as the *independent* variables. In cases where the precise form of this functional relationship is known, it is normally expressed as an equation.**

A function is generally defined only over some specified range of values of the independent variable(s). For example, in (3.1) it might be the case that economies of scale would cause the variable cost per car to drop if daily production exceeds 1000 cars, so that the coefficient of N (and possibly also the constant term) would change when N exceeds 1000. In that case (3.1) would only hold so long as daily output does not exceed 1000 cars; in other words, while N lies in the range 0 to 1000. We then say that the *domain* of the function in (3.1) is $0 \leq N \leq 1000$.

In general,

- **The range of values of the independent variable(s) over which a function is defined is called the *domain* of the function.**

Although in our definition of functions we have allowed for the possibility of more than one independent variable, much of our attention will be focused on functions of a single independent variable. Such functions can be expressed pictorially by means of *graphs*, which provide a way of obtaining a better understanding of functional relationships. The next section therefore discusses the basic principles of graph-plotting.

We then introduce some standard forms of function and their graphical representations. Although there is an infinite variety of possible functional forms, a few well-known types are sufficient to express, or at least approximate, most forms of relationship between variables which you are likely to encounter in practical applications. We therefore begin with an extensive discussion of *linear* functions of a single independent variable – that is, functions whose graphs are straight lines. We emphasise such functions for two reasons: first, they are of considerable practical importance; and second, they provide a simple vehicle for learning about graphical representation of functions. We then move on to consider some other important standard functions: the logarithmic, exponential and quadratic functions, and functions of the form a/x^n.

3.2 The Basics of Graph-plotting

3.2.1 *The Cartesian plane*

Relationships involving one dependent and one independent variable can be displayed effectively by means of a two-dimensional graph, using one dimension to represent each variable. Relationships involving larger numbers of variables would require more dimensions; it is just possible to visualise what a three-dimensional graph might look like, but a 'pictorial representation' in more than three dimensions defies the imagination! We shall therefore restrict our discussion to two-dimensional graphs.

Graphs in two dimensions are drawn on the *Cartesian plane*, and refer to a pair of *Cartesian axes*. These are two lines at right-angles, as shown in Figure 3.1. The horizontal axis is conventionally allocated to the independent variable, often denoted by x; this axis is then called the x-axis. Similarly, the vertical axis is allocated to the dependent variable, and frequently labelled as the y-axis. The point where the two axes meet usually corresponds to the point where both variables are zero, and is called the *origin*. Each axis is labelled with the values of the corresponding variable, starting at the origin. Values on the x-axis to the right of the origin are positive, those to the left negative. On the y-axis, values above the origin are positive and those below are negative.

There is no need to use the same scales on both axes. For

FIGURE 3.1
The Cartesian plane

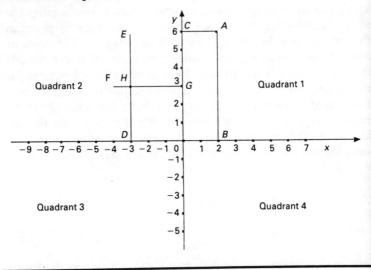

example, values on the x-axis may rise by steps of one and those on the y-axis by steps of one thousand. The only requirement is that equal distances on an axis must generally represent equal changes in the value of the variable concerned. One exception to this is that the scale on an axis may be 'broken' at some point, continuing with much higher values above the break point. This is done in order to cover a large range of values on a single axis; the break point should be clearly indicated by a gap or zigzag section on the axis.

By referring to the values marked on the axes, we are able to locate points in the Cartesian plane. To locate a point uniquely we need two values, one referring to each axis. The two values are called the *coordinates* of the point, and can be thought of as its 'address'. Let us first examine how we can determine the coordinates of a point such as point A in Figure 3.1, which is already plotted in the Cartesian plane.

The x-axis coordinate of A is obtained by drawing line AB from A parallel to the y-axis. The point B where this line meets the x-axis gives 2 as the x coordinate of A. Similarly, the y coordinate of A is obtained by drawing a line from A parallel to the x-axis

This is line AC, which gives 6 as the y coordinate of A. Thus the coordinates of A are $x = 2$ and $y = 6$, written in a shorthand format as $(2, 6)$. *Note that by convention the first co-ordinate here relates to the x-axis and the second to the y-axis.*

We can reverse this process if we wish to identify the point specified by a pair of coordinates. For example, to identify the point $(-3, 3)$, draw a line parallel to the y-axis through the point on the x-axis where $x = -3$. This gives line DE in Figure 3.1. Now draw a line parallel to the x-axis through the point $y = 3$ on the y-axis (line FG). The two lines meet at H, which is therefore the point with coordinates $(-3, 3)$. The lines used above to establish the correspondence between the coordinates and the point in the Cartesian plane are not normally drawn; we can usually plot a point or read its coordinates simply by inspection.

The axes divide the Cartesian plane into four parts, known as *quadrants*, (see Figure 3.1). Points in the first quadrant have both their coordinates positive, which is why the first quadrant is also known as the *positive quadrant*. Both coordinates of points in the third quadrant are negative, while points in the second and fourth quadrants have one positive and one negative coordinate.

3.3 Linear Functions of a Single Variable and their Graphs

3.3.1 *General form and its interpretation*

A linear function of one variable has the following general form:

$$y = f(x) = ax + b \qquad (3.2)$$

where a and b are constants.

Here y is a function of the single variable x. Moreover, x only occurs as x^1, and not squared, square-rooted, or raised to any other power. We will see later that this is characteristic of all linear functions.

We have already come across one example of a linear function in section 3.1. If you look at function (3.1) and compare it with the general form (3.2), you can see that (3.1) is a special case of (3.2) with $a = 6$ and $b = 1000$.

The constants a and b in a linear function have a practical interpretation, as we can see by examining the linear function (3.1) in more detail. Take a, the constant multiplying x, first. When N increases from 1 to 2, C will increase by 6; the same will be true if N increases from 1000 to 1001, or indeed for any unit increase in N. So the 6 in (3.1) may be interpreted as the *rate* at which C increases with N. Applying this interpretation to the general expression for a linear function in (3.2), we can say that

- **In the function $y = ax + b$, a is the *rate of change of y with x*, that is, the increase in the value of y as x increases by one unit. If a is negative, then y decreases with x.**

(You may have come across the 'rate of change' concept in economics under the alternative name of 'marginal': here the marginal cost of one car is 6, expressed in thousands of pounds.)

The interpretation of b in the general expression for a linear function in (3.2) is easier. All we need do is to note that when $x = 0$ the value of y is b. So we see that:

- **In the function $y = ax + b$, b stands for the value of y when x is zero.**

For example, we saw that in the linear function in (3.1) $b = 1000$. This is the value of the function (that is the total daily cost) when $N = 0$ and no cars are produced.

3.3.2 *The graph of a linear function*

We can now draw the graph of a linear function of a variable in the Cartesian plane, taking as an example our cost function from (3.1), $C = 6N + 1000$. But first we need to define just what we mean by the graph of a function:

- **The graph of a function is the set of points in the Cartesian plane whose coordinates satisfy the function (that is, make the two sides of the equation of the function equal).**

So, in order to plot the graph of a function, we need to identify some points in the plane whose coordinates satisfy the function (we cannot identify the complete set, of course, because generally it contains an infinite number of points!). Often the easiest way to do this is to substitute values of the independent variable into the equation of the function, and find the corresponding values of the

FIGURE 3.2
Graph of the function 6N + 1000

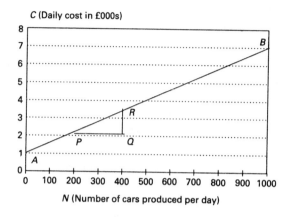

dependent variable. For example, in equation (3.1), when $N = 100$, $C = 1600$; when $N = 200$, $C = 2200$, and when $N = 1000$, $C = 7000$. The points with coordinates (100, 1600), (200, 2200) and (1000, 7000) will therefore all lie on the graph of the function $C = 6N + 1000$. We show below some more pairs of N and C values which belong to the graph:

N	0	100	200	400	600	800	1000
C	1000	1600	2200	3400	4600	5800	7000

Here we have chosen to include values of N from 0 to 1000 because we had already identified this as the domain of our function.

In Figure 3.2 the points which have these coordinates have been joined by the line AB. This line is the graph of the function $C = 6N + 1000$ for $0 \leq N \leq 1000$.

3.3.3 Features of the graph of a linear function of one variable

The most notable feature of the graph in Figure 3.2 is, of course, the fact that it is a straight line. If you look back to the table of

(N, C) coordinates we used to plot the graph, you can see why this is so: equal increases in N produce equal increases in C. Whenever N increases by 100, C increases by 600. This would *not* happen if the equation of the function contained powers other than N^1 (you can check this by examining what would happen if we had $C = 6N^2 + 1000$ instead). So we can say that:

- **Linear functions of a single variable have graphs which are straight lines.**

And conversely, that:

- **All straight lines represent linear functions.**

This, as we said earlier, is why such functions are called linear.

One advantage of recognising that a function to be plotted is linear is that we only actually need two points to be able to plot the graph. For example, in the case of the function $C = 6N + 1000$, once we knew that the points (200, 2200) and (1000, 7000) satisfied the equation of the function we could have joined them and then extrapolated the line obtained to get AB the graph of the function. If we were feeling cautious we might want to plot a third point, just to be on the safe side.

Next we need to relate the shape of the graph of the function to its equation. To investigate this, we will again use equation (3.1):

$$C = 6N + 1000.$$

We already know that the 6 here represents the rate of change of C with $N - C$ increases by 6 for every unit increase in N. In terms of the graph, that means that for every unit we move along parallel to the N-axis, the graph goes up by 6 units. We call this increase in C per unit increase in N the *slope* or *gradient* of the graph; it is the graphical equivalent of the rate of change of C with N. So our graph in Figure 3.2 has slope 6. More generally we can say:

- **The *slope* or *gradient* of the graph representing the linear function $y = ax + b$ is a, and is equal to increase in y divided by corresponding increase in x.**

The graph of a linear function is a straight line, of course, precisely *because* its slope is a constant. To verify this, and to see how we can compute the slope from the graph, consider the triangle PQR

in Figure 3.2. At P, $N = 200$ and $C = 2200$, while at R, N is 400 and C is 3400. So the slope of the line is $QR/PQ = (3400 - 2200)/(400 - 200) = 1200/200 = 6$, as already established. If you do the same calculation for any other pair of points on the graph, you will get exactly the same answer. This is in contrast to some of the non-linear functions we will meet later in the chapter, whose slopes vary from one point to another. Calculating their slopes is much more difficult, and indeed will occupy much of Chapter 5.

Now let us turn to the 1000 in our particular equation $C = 6N + 1000$. We have already seen that this is the value of C when $N = 0$. Looking at the graph it is clear that when $N = 0$ the graph crosses the vertical axis; this point is known as the *intercept* of the graph. In our example, then, the intercept is 1000.

We have to be a little bit careful about the general definition, because just occasionally a graph may be plotted with a horizontal axis which does *not* start from zero – perhaps because the domain of the function does not include zero. In such a case, the point where the graph crosses the vertical axis would *not* be equal to the value of y when $x = 0$. So we can only say that *if* the x-axis starts from zero, then:

• **The intercept of the graph of a linear function $y = ax + b$ is the point at which the graph crosses the y-axis, and has the value b.**

The two constants a and b in the equation $y = ax + b$ thus determine between them the position and orientation of the graph; by giving them different values we can obtain different lines in the Cartesian plane. We shall look into this in more detail in the next section.

3.4 More about slopes and intercepts

So far we have restricted our discussion of straight-line graphs to equation (3.1), in which both a, the slope, and b, the intercept, had positive values. But there is no reason why that must be so in general. Let us now see what happens if we have a negative value of a – that is, a negative slope. An example of this would be the equation $y = 3 - x$; if we rearrange this as $y = -x + 3$, and compare it with the general equation (3.2), we find that $a = -1$ and $b = 3$. The intercept, then, will be $+3$, but what does the slope of -1 mean?

FIGURE 3.3
Illustrating slopes of linear functions

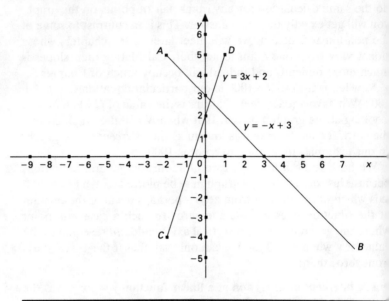

The graph of $y = -x + 3$ has been plotted as line AB in Figure 3.3. You can see that this time y *decreases* as x increases. This agrees with our method for calculating the slope, as can be verified by computing the slope at a point on AB. For example, at A we have $x = -2$, $y = 5$, and at $x = 0$ we have $y = 3$. Thus the slope is (change in y/corresponding change in x) = $(5 - 3)/(-2 - 0) = 2/-2 = -1$, as before. Roughly speaking, we can say that a graph with a negative slope goes 'downhill' from left to right.

In contrast, the slope of the function $y = 3x + 2$ (line CD in Figure 3.3) is $+3$. Hence we would expect y to increase as increases, and the graph shows that this is indeed so. A graph with positive slope goes 'uphill' from left to right. The absolute value of the slope indicates how steep the corresponding line is. The larger the absolute value of the slope, the steeper the line – assuming, of course, that the graphs in question are plotted on the same scales.

This is demonstrated in Figure 3.4. The slope of $y = 4x + 3$ is four times that of $y = x + 5$, so the graph of $y = 4x + 3$ rises much

FIGURE 3.4
The link between slope and steepness

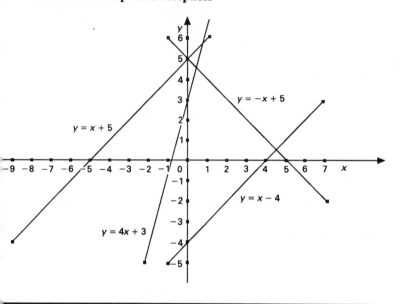

more steeply than that of $y = x + 5$. It follows from this discussion that two straight lines with the same slope are parallel. The lines $y = x - 4$ and $y = x + 5$ in Figure 3.4 demonstrate this – both have a slope of 1.

Another, slightly less obvious fact (which we shall not prove formally) is that if the product of the slopes of two lines is -1, then the lines are perpendicular to each other. Figure 3.4 demonstrates this; the lines $y = x - 4$ and $y = -x + 5$ have slopes of 1 and -1 respectively, so the product of their slopes is -1, and the lines are indeed at right-angles.

Let us now summarise what we have discovered so far about linear functions and their graphs.

1. The intercept of the graph of the linear function $y = ax + b$ is the value of the function when $x = 0$, and is equal to b. If the domain of the function includes $x = 0$, then the intercept is the point at which the graph of the function crosses the y-axis.

2. The rate of change of a linear function $y = ax + b$ is the increase in y per unit increase in x, and is equal to a. This is also equal to the slope or gradient of the graph of the function.
3. If the graph of a function has a positive (negative) slope, then y increases (decreases) as x increases; conversely, if a function has a positive (negative) rate of change so that y increases (decreases) as x increases, then its graph will have a positive (negative) slope.

To get a better feel for the practical implications of all this, we will look at a few more examples of functions and their graphs.

EXAMPLE 1

$$y = -1 - x.$$

Comparing this with the standard linear function $y = ax + b$, we can see that $a = -1$ and $b = -1$. The graph thus has a slope of -1 and an intercept of -1; it is shown in Figure 3.5, line (a).

EXAMPLE 2

$$y = 4x.$$

Here $a = 4$, and there seems to be no b-value – or at least, $b = 0$. This means that the graph has zero intercept, so that it passes through the origin, as seen in Figure 3.5, line (b).

EXAMPLE 3

$$y = 11.$$

This at first sight does not look like a linear function at all – it contains no x-term. However, it could be written as $y = 11 + 0x$, from which it is clear that its slope is zero. So as x increases, y does not change at all, giving a graph which is a horizontal line through the intercept at $y = 11$ (see Figure 3.5, line (c)).

EXAMPLE 4

$$x = 2.$$

FIGURE 3.5
Illustrations of linear graphs

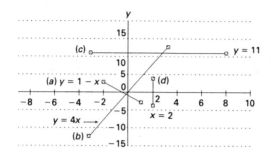

This really is an exceptional case – it is not a linear function at all, since it contains no y-term. Graphically it would be represented simply by a vertical line through $x = 2$, indicating that, whatever the value of y, x is always equal to 2. Figure 3.5, line (d), shows this.

3.3.5 *Business applications of linear functions and their graphs*

Having set up the machinery of linear functions and graphs, we can now use them to convey the essential features of relationships between variables arising in practical problems. In fact we have already seen one such example; in (3.1) we had a situation where the total daily cost of production at a car plant was expressed as a function of the number of cars produced. The function (3.1) is known as a *linear cost model*. This section illustrates some other important linear models that occur in business and economics.

3.3.6 *A linear demand function*

Our experience tells us that in general, when the price of a commodity is increased, the demand for that commodity will go down.

For example, consider a farmer who has free-range chickens, and sells their eggs by direct door-to-door delivery. He has noted that when he charges 8 pence per egg he sells 400 eggs daily, but if he charges 12 pence he sells only 300 eggs daily. If we assume that the number of eggs sold daily is a linear function of the price of each egg, then we would like to know the exact form taken by the equation of that function. We can then use the equation to estimate how many eggs will be sold daily if the price is set at, say, 9 pence per egg.

Let us use D to denote the number of eggs sold per day, and P to denote the price per egg in pence. Then, using the general form of a linear function (3.2), D in terms of P can be written as:

$$D = aP + b. \tag{3.3}$$

We need to determine the values of the constants a and b in (3.3), using the information given. We know that when $P = 8$, $D = 400$, and when $P = 12$, $D = 300$. This means that when the price P rises by 4 pence the number of eggs sold falls by 100 eggs. Therefore the *rate of change* of D with P is -100 eggs per 4 pence or -25 eggs per penny. This means that in (3.3) we have:

$$a = -25$$

since a is the rate of change of D with P, and P is in pence.

The other constant in (3.3), b represents the value of D when $P = 0$. b can now be readily computed by using the fact that when $P = 8$, $D = 400$, so that

$$400 = -25 \times 8 + b,$$

or

$$400 = b - 200.$$

The processes by which this equation can be *solved* to find the value of b will be covered formally in Chapter 4. However, with such simple numbers you can probably see by direct inspection that

$$b = 600.$$

FIGURE 3.6
Supply and demand functions

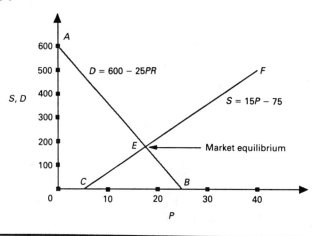

Of course, we could have used the values $P = 12$, $D = 300$ instead – we would have arrived at the same value for b.

We now have the complete function linking the number of eggs sold, D, and their price in pence, P; it is

$$D = 600 - 25P. \tag{3.4}$$

This is known as a *linear demand function*; its graph is line AB in Figure 3.6. If the price of eggs is set at 9 pence each then, either by using (3.4) or by reading from the graph, we would estimate the number of eggs sold daily to be

$$D = 600 - 25 \times 9 = 375 \text{ eggs.}$$

We can easily see from Figure 3.6, and confirm using (3.4), that when $P = 24$, $D = 0$. This implies that if the farmer charges 24 pence each for eggs, he will sell none. On the other hand, we can also see that when $P = 0$, $D = 600$. That is, if the farmer gives away the eggs free then 600 eggs will be taken. These conclusions are clearly not very sensible – there is probably always going to be a small demand for eggs however high the price. Part of the problem is that $P = 0$ and $P = 24$ are well outside the range of observed values of $P = 8$ and $P = 12$ which were used to estimate the equation of the function. Using a concept we introduced earlier, we can say that the domain of the function as we have derived it probably does not include $P = 0$ and $P = 24$.

Generally we must always be clear as to the domain of any function whose equation we have estimated. For example, in (3.4) $P = 25$ is certainly not in the domain of the function as it would imply $D = -25$. That is, if the price is 25 pence the farmer buys 25 eggs rather than selling any! We must always be wary of extrapolating the domain of the function beyond the range of values of the independent variable(s) used to derive the equation of the function.

3.3.7 A linear supply function

A *demand function* such as that in (3.4) is used in economics to link the price charged for a good or service to the number of units of that good or service which customers would be willing to purchase at that price. In a similar way, *supply functions* are used by economists to link the quantity of a good or service that suppliers will provide to the price which they can obtain for it. The *market equilibrium* for the good or service is reached at the price where supply equals demand, as the next example will illustrate.

Suppose that the farmer in the last example can sell in bulk to supermarkets, but only at a very low price of 5 pence per egg. Therefore he is not willing to sell any eggs door-to-door at 5 pence or less, since he incurs lower costs in selling to supermarkets. On the other hand, for every extra penny he can obtain on the price of door-to-door eggs, he is willing to divert an additional 15 eggs from supermarket to door-to-door sales.

We shall derive a function linking the number of eggs the farmer is willing to sell door-to-door to the price he can obtain for such eggs, assuming the function to be linear. We can then use this function to estimate the price of door-to-door eggs in the market in which the farmer operates.

Let S be the number of eggs per day which the farmer is willing to supply door-to-door, and P be the price in pence at which they sell. If a linear function links S and P then, using the general form (3.2), we can write:

$$S = aP + b \qquad (3.5)$$

We know that the farmer will increase the supply of eggs by 15 per penny rise in their price P. So the rate of change of S with P is $+15$, and thus in (3.5) we have

$$a = 15.$$

We also know that the farmer will supply no eggs at 5 pence. Hence in (3.5)

$$0 = 5a + b$$

or

$$0 = 5 \times 15 + b$$

or

$$0 = 75 + b.$$

Since b added to 75 equals zero, the only possible value which b can take is -75; so

$$b = -75.$$

Thus (3.5) becomes

$$S = 15P - 75. \tag{3.6}$$

This is known as a *linear supply function*; its graph is line CD in Figure 3.6.

In order to estimate the price of door-to-door eggs in the market where the farmer operates, we note that demand falls as their price rises, while supply falls if price goes down. A stable price is that which makes the quantity of eggs the farmer is willing to supply equal to the quantity people would demand. In terms of Figure 3.6 this stable price is the one at which the graphs of the supply and demand functions intersect – point E. If the graphs are plotted accurately we can read off the price P^* (in pence) corresponding to point E. P^* is 16.875, and the corresponding quantity supplied and demanded is 178.125.

That is, the farmer should charge 16.875 pence per egg, or £1.35

for every 8 eggs. This is called the *market equilibrium price* as the farmer will be willing to supply exactly the quantity of eggs that people are willing to buy at this price. That quantity can be read as the S or D value corresponding to E in Figure 3.6 and it is 178.125 eggs per day – not a very realistic proposition! As often happens, the mathematical solution to the problem is not realistic in practice; the farmer would actually have to sell 178 eggs per day, or else adjust the price slightly. We will see in Chapter 6 that there is often a difference of this kind between the mathematical 'best' solution and what can be achieved in practice.

Instead of solving the problem graphically, we could have used (3.4) to estimate the quantity people would demand at price P^* as

$$D = 600 - 25P^*.$$

Similarly, using (3.6) we could have estimated the quantity the farmer would be willing to supply at price P^* to be

$$S = 15P^* - 75.$$

If the price is to be stable we need supply to just meet demand, so that

$$D = S, \text{ or } 600 - 25P^* = 15P^* - 75.$$

You can verify that the value

$$P^* = 16.875,$$

which we found graphically, makes the two sides of the equation equal. What we are really doing here is solving the demand and supply equations *simultaneously*, to find a point which satisfies both. As we have seen, this corresponds to the point where the graphs of the two equations cross. We will be returning to this topic of simultaneous equations and their solution in Chapter 4.

You will find some more practical applications of linear functions in the exercises following this section, and in Chapter 6.

Exercises 3.1

1. Plot the graph of the function $y = 6 + 2x$ for $0 \leq x \leq 5$.
2. Plot the graph of the function $y = 2 - 3x$ for $-2 \leq x \leq 2$.
3. A car of a particular model is estimated to lose 20% of its initial purchase value per annum. In addition, a cost of 20 pence is incurred in running expenses for every mile the car is driven.
   (a) Using P for the purchase price of the car in £s write its value as a function of t, the number of years since the purchase of the car.
   (b) Write the total cost in lost value and running expenses during the third year of the life of the car as a function of the number of miles it is driven during that year.
   (c) Suggest sensible domains for the two functions you have written.
4. In the functions that follow treat y as the dependent variable. In each case state:
   (i) the y-intercept;
   (ii) the slope;
   (iii) the rate of change of y with x; and
   (iv) which pairs of graphs (if any) would be parallel or perpendicular.
   (a) $3y = 6x - 6$;   (b) $y - 2 = 3x + 4$;
   (c) $2(y + x) = 3$;   (d) $y/5 - 3 = 2x$;
   (e) $-6y + 2 = 2x - 4$.
5. A promoter wants to print a number of leaflets for distribution to potential customers and is told by a printer that each leaflet will cost 10 pence but that there is a minimum charge of £100. Write the cost C to the promoter as a function of the number of leaflets printed.
6. The printer in question 5 notes that material and time to set up a printing run cost £500 for printing 10 000 leaflets and £900 for printing 25 000 leaflets. If material and set-up costs are a linear function of the number of leaflets printed write down this function.
7. In the case of question 6, what is the fixed cost and what is the additional cost of printing one leaflet? How many leaflets need to be printed in a run for the printer to break even on the run?

3.4 Some Nonlinear Functions

3.4.1 *Quadratic functions*

Linear functions are arguably some of the most widely used; their graphs are easy to draw, to recognise and to understand, and they often give acceptable approximations to relationships between quantities. On many occasions, however, the relationship between quantities cannot be approximated satisfactorily by a linear function. An example is the interest earned from an amount deposited in a bank. Even if interest rates stay constant, as the amount builds up through interest added, the interest earned each year will not be a linear function of the initial amount deposited, as we will see later in this chapter. The rest of the chapter introduces some of the more important nonlinear functions, beginning with quadratic functions which in a sense are the 'next most simple' form after linear functions.

A quadratic function has the general form:

$$y = ax^2 + bx + c \tag{3.7}$$
where $a \neq 0$.

Thus a quadratic function in its most general form contains an x-squared term as well as an x-term and a constant. The quadratic function, with its x-squared term, will *not* result in a straight line when plotted. In fact the graph of a quadratic function gives a shape known as a *parabola*; some typical parabolic graphs are shown in Figure 3.7.

A parabola has just one turning point, which can be either a lowest point, known as a *local minimum*, or a highest point known as a *local maximum* (see Figure 3.7). (The significance of the word local will become apparent in Chapter 5.) Whether a particular quadratic function leads to a maximum or minimum depends on the sign of the constant a in (3.7). A local minimum point is obtained when a in (3.7) is positive. For example, parabolas (a) and (c) in Figure 3.7 are of this type. A negative a in (3.7) would mean that the corresponding graph has a local maximum point, like parabolas (b) and (d) in Figure 3.7. The value of the independent variable x which corresponds to the local maximum or local minimum value of the function is often a matter of special interest; for instance, if parabola (d) represented profit as a func

FIGURE 3.7
Some typical parabolas

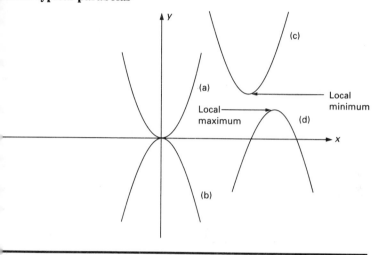

tion of output, then we would be interested to know what level of output led to profit being maximised.

The other constants b and c in expression (3.7) determine the position of the parabola; (a) and (b) in Figure 3.7 have their turning point at the origin because $b = c = 0$, while (c) and (d) result from non-zero values of b and c.

The following example will give you more insight into the workings of quadratic functions.

EXAMPLE

Plot the graph of the function $y = x^2 - 4x + 2$ for $-2 \leq x \leq 4$. Does the graph show a local maximum or a local minimum point within the range of x-values plotted? If so, what is the value of x at the maximum or minimum, and what is the corresponding value of the function?

SOLUTION

We need a whole set of points to be able to plot a quadratic, since its graph is a curve. We obtain these by taking a series

FIGURE 3.8
Graph of the function $y = x^2 - 4x + 2$

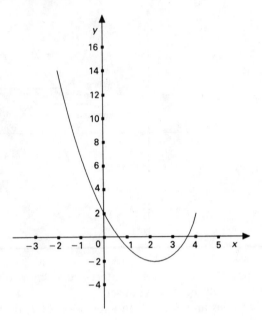

of values of x and finding the corresponding values of y. For example, when $x = 1$, $y = 1 - 4 + 2 = -1$. In a similar way, we find that the following pairs of x and y values satisfy the function:

x	-2	-1	0	1	2	3	4
y	14	7	2	-1	-2	-1	2

Figure 3.8 shows the curve obtained by joining the points with the above coordinates. This curve is therefore the graph of the function $y = x^2 - 4x + 2$. The parabola plotted seems to show a local minimum value at the point where $x = 2$. The value of the function at that point is -2. Generally speaking, we can only get an estimate of the position of the minimum

(or maximum) from the graph; to find its position more precisely, we would need to evaluate the function close to, and on both sides of, $x = 2$ to verify that the function does not take any lower value than -2. In this case, such calculations confirm that at $x = 2$ we do indeed have the lowest value for the function in the locality. Much of Chapter 5 will be devoted to discussion of a more direct method for finding the maximum or minimum of a function, without the need to plot graphs or calculate lots of values.

Quadratic functions are closely related to quadratic equations, which we will be discussing in Chapter 4.

3.4.2 Exponential functions

These are functions where the independent variable appears as an exponent or power. Their general form is:

$$y = ab^x \qquad (3.8)$$
where a and b are constants, with $b > 0$.

Functions of this kind occur frequently in business applications. They represent situations where the rate of growth of a quantity is proportional to its current value.

EXAMPLE

Consider the case where an amount of £1000 is invested at an annual interest rate of 10%, compounded annually. What is the value of the investment after 5 years?

SOLUTION

After one year the value of the investment is £V_1, where:

$$V_1 = (1 + 0.1) \times 1000.$$

After two years the value of the investment is £V_2, where:

$$V_2 = (1 + 0.1)V_1 = (1 + 0.1)[(1 + 0.1) \times 1000] = (1 + 0.1)^2 \times 1000 = 1.1^2 \times 1000.$$

FIGURE 3.9
Graph of the function $y = ab^x$, $b > 1$

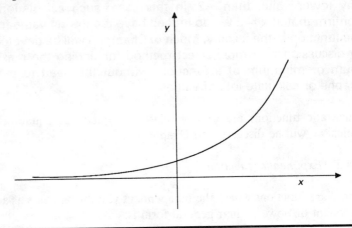

By repeated application of this approach it can be seen that after 5 years the value of the investment will be £V5, where:

$$V5 = 1.1^5 \times 1000.$$

It is easy to generalise this expression to find that after t years the value of the investment is:

$$V(t) = 1.1^t \times 1000.$$

By comparing this function with expression (3.8), you can see that $V(t)$ is an exponential function with $a = 1000$ and $b = 1.1$. This indicates that the rate at which the investment increases in a year is proportional to the size of the investment at the start of that year.

Figure 3.9 shows the general shape of the graph of an exponential function, when $b > 1$. As you can see from the graph, the rate of increase of y with x is initially slow, getting larger the larger y becomes. It is this case that has given rise to the term 'exponential growth' in common usage. For instance, where the growth of human or animal populations is not controlled, exponential growth is typically obtained: the larger the population, the larger the rate of growth.

The number e, which has the approximate value 2.718 and features in many mathematical formulae, was briefly mentioned in Chapter 1. The function $y = e^x$ is known as the *natural* exponential function (because e arises 'naturally' as the result of processes such as finding the sum of the series:

$$1/0! + 1/1! + 1/2! + 1/3! + 1/4! + \ldots ,$$

as you can verify by working out this sum numerically.) In fact, e^x is so common that it is often referred to simply as *the* exponential function. One feature of the natural exponential function, as we will see in Chapter 5, is that its rate of change is the same as the value of the function itself. The exponential function $y = e^x$ is so important that its values are tabulated in books of mathematical tables; these days, they can also be obtained from many pocket calculators.

Suppose, for example, that we want to plot the graph of $y = e^x$ for $0 \leq x \leq 3$. The following pairs of (x, y) values satisfying the above function can be found either from tables or with a calculator:

x	0	0.50	1.000	1.50	2.00	2.50	3.0
y	1	1.64	2.71	4.48	7.39	12.18	20.1

Figure 3.10 shows the graph defined by these points.

An exponential function does not always show growth. We can also have exponential decline. In this case in the general expression $y = ab^x$ for an exponential function we have $b < 1$. To see which way exponential decline works let us plot the function $y = 10 \times 0.4^x$ where $0 \leq x \leq 3$. The table below shows some (x, y) values satisfying the function $y = 10 \times 0.4^x$, and Figure 3.11 shows the graph of the function.

x	0	0.500	1	1.500	2.0	2.500	3.0
y	10	6.324	4	2.529	1.6	1.012	0.6

FIGURE 3.10
Graph of $y = e^x$ for $0 \leq x \leq 3$

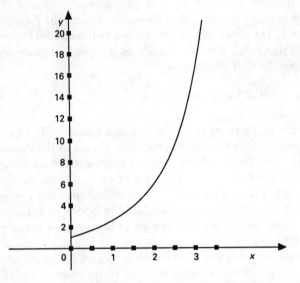

You can see in Figure 3.11 an illustration of exponential decline. The rate of decline is fast initially and slows down as the value of the function approaches zero.

3.4.3 Inverse functions

In Chapter 1 we came across the idea of inverse operations – for instance, we said that division was the inverse of multiplication, since a division will 'undo' or reverse the effect of a multiplication. In a very similar way we can define *inverse functions*. It is probably easiest to illustrate the concept by an example.

Consider the linear function $y = 2x$, from which, given any x-value, we can calculate the corresponding value of y: when $x = 1$, $y = 2$ and so on. Now imagine that we want to answer the question: given the value of y, is there a function which will enable us to get back to the corresponding value of x? In this case, to get from x to y, we multiplied x by two; so to get back from y to x, we

FIGURE 3.11
Graph of 10×0.4^x for $0 \leq x \leq 3$

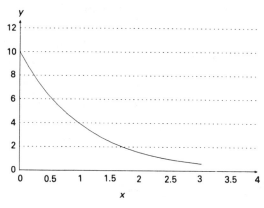

need to divide y by two. The function expressing this relationship is therefore $x = y/2$, and this is the inverse function of $y = 2x$.

You can see that what we have done is to 'change the equation round' to make x the dependent variable and y the independent variable. In Chapter 4 we will be exploring formal algebraic manipulations to enable us to do this 'changing round' more easily; for the time being, the important point is the idea of being able to invert a function. In fact we can say that $y = 2x$ and $x = y/2$ are each the inverse of the other, since we can use them in either direction to move from x to y or vice versa. We are really just expressing the same relationship in two different ways.

Not all functions have inverses; for example, if $y = x^2$ then x could be equal to the positive *or* the negative square root of y. Thus we can say that when $x = -3$, $y = 9$, but there is no function which will get us back from $y = 9$ to $x = -3$. There will always be an ambiguity about the sign; $y = x^2$ thus does not have an inverse function.

3.4.4 *Logarithmic functions*

The logarithmic function is the inverse of the exponential function.

Take, for example, the exponential function $y = 10^x$. If $x = 2$ we can compute that $y = 100$. Suppose we were to 'reverse' the question and ask:

If $y = 100$, what does x have to be to satisfy $y = 10^x$?

The answer is pretty obvious: x has to be 2. Another way of expressing this is to say that

- **2 is the power to which 10 must be raised to give 100.**

The formal way of describing this relationship is to say:

- **2 is the logarithm of 100 to base 10.**

This is written as:

$$\log_{10} 100 = 2.$$

Similarly, $\log_{10} 1000 = 3$ because raising 10 to the power 3 gives 1000. We can extend this process to any base.

For example, $\log_2 8 = 3$ (because $2^3 = 8$), $\log_3 81 = 4$, and so on.

In general, if $a^x = y$, we say x is the *logarithm* of y to the base a, and write:

$$x = \log_a y, \qquad (3.9)$$

if it is the case that:

$$a^x = y.$$

A couple more examples illustrate this definition: $\log_3 27 = 3$ because $3^3 = 27$, and $\log_{10} 10\,000 = 4$ because $10^4 = 10\,000$.

By far the most common bases for logarithms are 10 and $e = 2.718$. Logarithms – or logs, as they are often abbreviated – to a base of 10 are called *common logarithms*; those where the base is e are called *natural logarithms*. Published tables of both these types of logarithm are available, and they can also be computed on many pocket calculators. In terms of inverse functions, the natural logarithmic function $x = \log_e y$ is the inverse of the natural exponential function $y = e^x$.

The base is often omitted in writing logarithms. Instead, we use the convention that if the logarithm is written **log** y, the base is 10,

FIGURE 3.12
Graph of $y = \ln x$ for $0 \leq x \leq 3$

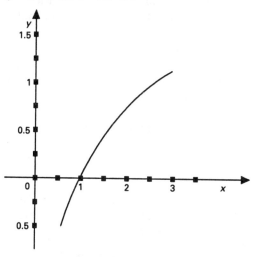

while if it is written **ln** y it is a natural logarithm.

Certain useful properties of logarithms follow from the way they are defined; these properties hold whatever the base of the logarithm.

- (i) **The logarithm of a number less than or equal to zero is not defined.**

It is easy to see why this should be so: if $y < 0$ and x were to be its logarithm then we would need $y = a^x$, where $a > 0$. But it is not possible for a^x to be negative as long as a is a positive number, no matter what the value of x. So we can find no value of x which will be the logarithm of a negative number y. The graph in Figure 3.12 illustrates this point, since it exists only where x is positive (note that we have now reverted to the more usual convention of writing y as the dependent variable). In the terminology introduced earlier, we could say that the domain of the log function is $x > 0$.

88 *Essential Mathematics*

- (ii) **The logarithm of the product of two numbers is the sum of the logarithms of the individual numbers.**

This rule can be derived from the fact that when we multiply powers, we add the exponents. The rule in mathematical terms is

$$\log(pq) = \log p + \log q.$$

This rule can be extended to the product of more than two numbers. An example will make this clearer:

$$\log(10 \times 10) = \log 10 + \log 10 = 1 + 1 = 2.$$

But $\log(10 \times 10) = \log(100) = 2$ as already shown, so the result clearly works in this case.

- (iii) **The logarithm of the quotient of two numbers is the difference between the logarithm of the numerator and the logarithm of the denominator.**

This rule follows from the rule for dividing powers. In mathematical terms the rule is:

$$\log \frac{p}{q} = \log p - \log q.$$

For example,

$$\log 1000/10 = \log 1000 - \log 10 = 3 - 1 = 2.$$

But $\log(1000/10) = \log(100) = 2$, so the result is correct.

- (iv) **$\log a^n = n \log a.$**

This property can be easily deduced from property (ii) if a^n is written:

$$a \times a \times a. \ldots$$

Finally, from the definition of logarithms one can readily deduce that:

- **The logarithm of 1 is zero whatever the base, because $a^0 = 1$ for any a.**
- **The logarithm of any number a to the base a is 1, because $a^1 = a$.**
- **The logarithm of any positive number less than 1 is negative.**

Figure 3.12 shows the shape of the graph of a logarithmic function when the base of the logarithm is greater than 1. You should compare this graph with the exponential function in Figure 3.10, to see how the shapes of the inverse functions relate to each other.

3.4.5 Semi-log graphs

From property (ii) of logs as noted above, we can deduce that $\log(ax) = \log a + \log x$. This result means that, whenever we multiply a number x by a *factor* a, we increase its logarithm by a *fixed* amount $\log a$. This fact can be exploited in plotting nonlinear functions so that straight line graphs can be obtained.

For example, earlier in this section we discussed the compound interest problem, where an amount of £1000 was invested at an annual interest rate of 10%, compounded annually. The increase in the amount invested during the first year will, of course, be £100; during the second year it will be 10% of £1100, or £110, during the third £121, and so on. The *amount* of the increase each year gets bigger as time goes on, so that plotting the value of the investment against time on an ordinary graph would give a curve rather than a straight line.

However, the *proportion* by which the investment increases each year is 10% or 0.1 of the amount in the account at the start of that year, and this proportion is constant. Therefore the *logarithm* of the amount invested will increase each year by the same amount.

We can see this by utilising the calculation performed earlier. We saw on page 81 that the value of the investment after 1 year, $V1$, was 1.1×1000, so that:

$$\log V1 = \log (1.1 \times 1000) = \log 1.1 + \log 1000,$$

using rule (ii) above. Similarly, we saw that $V2 = 1.1^2 \times 1000$, so that:

$$\log V2 = \log (1.1^2 \times 1000) = 2 \log 1.1 + \log 1000,$$

using rules (ii) and (iv).

More generally, we found that $V(t)$, the amount in the account after t years, was given by:

$$V(t) = 1.1^t \times 1000,$$

so that

$$\log V(t) = t \log 1.1 + \log 1000. \tag{3.10}$$

Now the log 1.1 and the log 1000 here are just constants, so you should recognise (3.10) as expressing the fact that $\log V(t)$ is a linear function of t. Thus in this situation if we plot, not the amount in the account, but its *log*, against time we will get a straight line.

This plotting would be very tedious if we actually had to find all the logs from tables or a calculator. However, special graph paper is available with a *logarithmic scale* already built in, and many computer packages will also plot such scales. An example of a graph using such a scale is shown in fig. 3.13. You can see that the scale is non-linear, as you would expect; and if you care to take a ruler to it you can verify that any two pairs of points in the same ratio (e.g. 100 to 1000 and 1000 to 10 000) are the same distance apart.

The other axis of the graph (that is, the t-axis in Figure 3.13) has an ordinary linear scale. The fact that only one of the axes is logarithmic gives rise to the name *semi-log* for this kind of graph. As you can see in the figure, when the 'steady rate of increase' data is plotted in this way, the resulting graph is indeed a straight line.

Semi-log graphs of this kind are therefore used when we wish to investigate whether figures show an approximately steady rate of increase or decrease; if their graph on semi-log scales is roughly a straight line, the steady rate of change is confirmed. The method can also be used for comparing different rates of change – for example, the growth rates of different companies. The steeper the line, the faster the rate of growth (or decline).

We can actually go a bit further than this. If you look back to equation (3.10), and compare it with the standard form of the

FIGURE 3.13
Semi-log graph of interest data

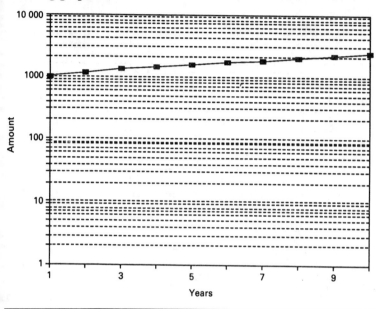

linear equation $y = ax + b$, you will see that a (the slope) corresponds to log 1.1, while b (the intercept) is log 1000. We can thus say that the slope of the straight line on the semi-log plot gives the log of the annual growth factor. Of course, in this case we already knew what that was, but in a case where we simply have annual figures, and want to estimate the growth rate, the semi-log plot can be an easy way of doing so – though you need to be careful in reading off intermediate values from the vertical axis, remembering that it is not linear.

There is a secondary use for such graphs. You can see from the scale in Figure 3.13 that a very large range is covered – from 1 to 10 000 – in such a way that both small and large figures can be shown in detail. Log scales are therefore often used when a wide range of data has to be covered in this way; you may encounter their use for this purpose in government publications such as *Economic Trends*.

FIGURE 3.14
Log-log graph showing $y = 10x^2$

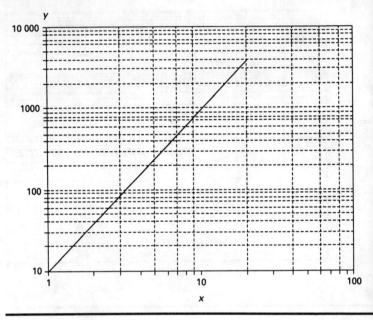

3.4.6 *Log-log graphs*

If we have a relationship of the form $y = ax^b$, the semi-log graph will not serve to produce a straight line. Instead, we need to observe that if we take logarithms of both sides of the equation, we get $\log y = \log (ax^b) = \log a + b \log x$. So if we set $\log x = u$ and $\log y = v$, we have $v = \log a + bu$, which you should recognise as the equation of a straight line (remember, $\log a$ and b are just constants).

So plotting u against v – or $\log x$ against $\log y$ – will give a straight line. As with semi-log graphs, to do this most simply we resort to graph paper with the log scales built in – *log-log* paper, as it is called – or to a suitable computer package. Fig. 3.14 shows such a graph, with the line representing the equation $y = 10x^2$ plotted.

To summarise what has been said about log scale graphs:

- Use a semi-log graph to produce a straight line from a steady rate of growth:
- Use a log-log graph to produce a straight line from a function of the form $y = ax^b$.

3.4.7 Functions of the form a/x^n

This is a function where the value of the dependent variable is inversely proportional to a power of the independent variable (not to be confused with an inverse function!).

The simplest function of this type has the form:

$$y = \frac{a}{x} \qquad (3.11)$$

where a is a constant and x is the independent variable. Notice that if we try to evaluate this function at $x = 0$, we will be dividing by zero, an operation which, as you learned in Chapter 1, is impossible! So we say that the value of the function *is not defined when $x = 0$*.

Such functions often arise in real life. For example, if the fixed costs of producing a certain product are £4000 per week, and x items per week are produced, then the share of fixed cost allocated to each item will be $4000/x$: that is, fixed cost allocated and number of items are inversely related.

Fig. 3.15 shows the shape of the graph of the function:

$$y = \frac{a}{x}$$

when $a > 0$. (The graph for $a < 0$ would exhibit similar features but in different quadrants). You can see that as x becomes a very small positive number, y becomes a very large positive number. For $x = 0$ the function is not defined, as we saw earlier. When x has a small absolute value and is negative the graph reappears at a large negative value of y.

As x becomes a very large positive number, y approaches zero, though it never actually becomes equal to zero. Similarly, when x is negative with a very large absolute value, y becomes a negative number with a very small absolute value. But y never actually

FIGURE 3.15
Graph of the function $y = a/x$, $a > 0$

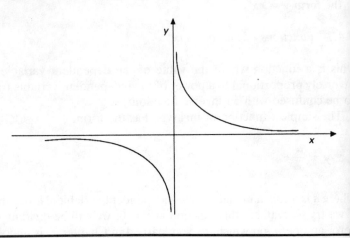

makes it to zero, though it gets closer and closer the larger the absolute value of x becomes. Hence the curve gets ever closer to the x-axis, as shown in Figure 3.15, but never actually touches it. The same happens with the y-axis for very small absolute values of x. Such a graph is said to be *asymptotic* to the axes, and in this case the x- and y- axes are the *asymptotes*, (i.e. the lines the curve is getting ever closer to). In other cases different asymptotes may apply (think about $1/(x - 2)$, for example).

As already noted, this function is not defined for $x = 0$. Such a function is said to be *discontinuous*. In this case the discontinuity is at $x = 0$. We can also say that the domain of the function does not include the point $x = 0$.

Graphs looking very similar to this one for a/x are obtained from other functions such as a/x^2, etc. The general name for a graph of this type, with two asymptotes, is a *hyperbola*.

As pointed out earlier, our aim in this section has been to introduce you to some of the more important nonlinear functions. The same kind of thinking should enable you to get to grips with other types of nonlinear function which you may encounter, as long as there is only one independent variable.

Exercises 3.2

(i) Plot the graph of each of the following functions, for $-4 \leqslant x \leqslant 4$. In each case name the type of curve obtained.
  (a) $y = 3x^2 - 10$; (b) $y = 10e^x$; (c) $y = 100/(x - 1)$;
  (d) $y = x^2 - x + 3$.
(ii) Plot $y = \log 100 x$ for $0 < x \leqslant 4$.

A newspaper publisher estimates that 500 000 copies of his paper will be sold daily if it is priced at £0.65 a copy. He estimates that if the price charged is £(0.65 + P) the number of copies sold will be (500 − 600 P) thousand.
(i) Write the publisher's revenue from the paper as a function of the change £P in the price. (Remember that revenue = price × quantity sold.)
(ii) Plot the function in (i) for $0 \leqslant P \leqslant 0.65$. What is the type of curve obtained?
(iii) Comment on what happens to revenue as the price charged per copy rises from £0.65 to £1.

A self-employed businesswoman purchases a car for the needs of her business, paying £20 000. She is given an annual tax allowance in respect of the car which is equal to 25% of the value of the car at the start of the year. The value of the car, for tax purposes, is deemed to be its initial purchase price minus total tax allowances to date.
(i) Compute the tax allowances the car generates at the end of each of the first five years of its life.
(ii) Plot the tax allowances in (i) against the age t of the car. Comment on the type of curve obtained.
(iii) Repeat the plot from (ii) using semi-log scales. Comment on the graph obtained and estimate from it the rate of depreciation of the car for tax purposes.

Expand the expression $\log((a^3 b^2 c^3)^{1/3} / a^2)$.

A car tyre manufacturer tests tyres under laboratory conditions to estimate their life in miles. He finds that the more accurately he wants to estimate the life of tyres, the more tyres he needs to test to destruction and so the higher the cost. The laboratory statistician has developed a formula which says that if the manufacturer wishes to estimate the life of his tyres to within $\pm d$ miles the costs, in £m, would be $540/d^2$. Sketch a curve of the cost against the precision with which the manufacturer estimates the life of the tyres. What practical conclusions can you draw from the curve?

6. Plot $y = 10^x$, $0 \leq x \leq 3$ on linear and on semi-log scales. Comment on the graphs obtained.

7. The following are annual average values of the Retail Price Index for the years 1987 to 1992 (based on 13 January 1987 = 100). Plot this series using a logarithmic scale. How would you describe the rate of inflation over this period?

Year	RPI
87	101.9
88	106.9
89	115.2
90	126.1
91	133.5
92	136.5

4 Equations

4.1 Introduction

Equations, as we saw in Chapter 3, are algebraic expressions involving the equals [=] sign. For example:

$$35 - x = 2x - 10$$
$$y = 4x - 6$$

are equations.

The [=] sign separates the equation into two sides: a left-hand and a right-hand side. The interpretation of the equation is that the expression on the left-hand side (LHS) is equal to the expression on the right-hand side (RHS).

Equations are extremely useful because they enable us to express certain requirements in decision situations which we want to analyse. An example will illustrate this.

EXAMPLE

A farmer has produced 10 000 kg of wheat. He may store the wheat produced in order to sell it when the price is more favourable. However, he estimates that the wheat loses 1000 kg through evaporation for every month that it is stored. The farmer has contracted to deliver 8000 kg of wheat. What is the longest time he can wait before he delivers the wheat?

SOLUTION

You can, of course, see the answer in this simple case without resorting to formal equations, but it is worth seeing how an equation can give the answer.

Let M be the number of months for which the farmer stores the wheat. The weight lost will be 1000 M kg and the wheat left will weigh 10 000 − 1000 M kg.

Thus if we use W for the weight of wheat left after M months, the functional relationship between W and M is:

$$W = 10\ 000 - 1000\ M \tag{4.1}$$

where W is in kg.

If the farmer must deliver 8 000 kg then we need to set $W = 8000$ and (4.1) reduces to:

$$10\ 000 - 1000\ M = 8000. \tag{4.2}$$

Expressions (4.1) and (4.2) are equations. The first is a function stating the *correspondence* between the weight of the wheat left (W) and the months it has been stored (M). For example $M = 1$ corresponds to $W = 9000$ kg and so on. The second is an equation defining the values of M that correspond to $W = 8000$ in the function (4.1).

Both (4.1) and (4.2) are linear equations. As in the case of linear functions,

- **A linear equation is one where no variable is raised to any power other than 1.**

We noted earlier that equation (4.2) results from wanting to find in (4.1) the value of M which corresponds to $W = 8000$ kg. This value of M makes the left- and right-hand sides of (4.2) equal and it is called the *solution* or *root* of equation (4.2).

- **Solving an equation is equivalent to finding the root or roots of the equation.**

Note that roots of equations have nothing to do with the square, cube and other roots we met when we dealt with powers.

As you might have gathered from the discussion above there are many kinds of equation: those that involve one variable, many variables, products of variables, various powers of variables and so on. An equation may have no (real) root, one root or many roots. The next two sections cover specific procedures for solving linear

FIGURE 4.1
Wheat available after M months

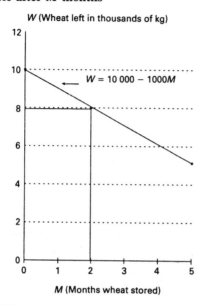

and quadratic equations involving one variable. Then two sections follow introducing procedures for dealing with a single equation involving more than one variable, and for sets of equations involving many variables. The chapter ends with an introduction to the solution of linear inequalities involving one variable.

4.2 Solving Linear Equations Involving One Variable

(a) The graphical method

We can use equation (4.2) above to illustrate how linear equations with one variable can be solved.

One way to solve equation (4.2) is to remember that it defines the values of M that correspond to $W = 8000$ in the function (4.1). So to solve (4.2) we can plot the graph of (4.1) and from it find the values of M that correspond to $W = 8000$. The graph appears in Figure 4.1. Only one value of M corresponds to $W = 8000$ in the

FIGURE 4.2
Graph of $y = 5x - 15$

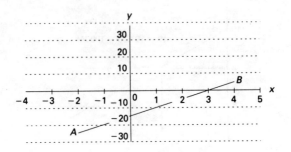

function (4.1). It is $M = 2$ and this is the solution to equation (4.2). It is no coincidence that equation (4.2) had a single root.

- **A linear equation of the form $ax + b = 0$ will always have just one root.**

In this case, we began with the function $W = 10\,000 - 1000\,M$, which indicated what we should plot. However, even when we are starting directly from an equation with only one variable, we can still use the graphical method, as the following example will illustrate.

EXAMPLE

Solve the equation:

$$5x - 15 = 0. \tag{4.3}$$

SOLUTION

We can think of this equation as defining the values of x that correspond to $y = 0$ in the function:

$$y = 5x - 15. \tag{4.4}$$

We can always link an equation to a function in this way so that an equation of the form $ax + b = 0$ defines the values of x that correspond to $y = 0$ in the function $y = ax + b$. (We will see later how we can manipulate equations if necessary to get all terms involving the independent variable on the same side of the equation.)

Once we link equation (4.3) to function (4.4), we can solve it by finding the value of x corresponding to $y = 0$ in the graph of the function (4.4). Figure 4.2 shows this graph. The value of x in Figure 4.2 corresponding to $y = 0$ is the point where the line AB crosses the x-axis. This point is $x = 3$. So the root of equation (4.3) is $x = 3$.

The graphical approach to solving equations is simple but time-consuming. Moreover, the roots obtained may not be very accurate unless high-precision graphs are plotted. What we need is an efficient, accurate and general method for solving equations. One such method is outlined next: it relies on algebraic manipulations rather than graphs.

(b) The algebraic method

The algebraic manipulations used to solve linear, or indeed non-linear, equations rely on one key rule. That is,

- **If we perform an algebraic operation on the LHS of an equation we must also perform the same operation on its RHS.**

This rule maintains the 'balance' between the two sides of the equation. At present we will exclude from this rule operations with zero; we will see later why this must be.

In order to illustrate how this rule, using the appropriate manipulations, can be used to solve a linear equation with one variable, we shall again use equation (4.2):

$$10\,000 - 1000\,M = 8000. \quad (4.5)$$

Our first aim is to separate terms involving the unknown quantity M from terms which are known. So we need to eliminate the 10 000 from the LHS of the equation. This we can do by subtracting 10 000 from the LHS. However, as we saw in the rule for operating on equations, we must then also subtract 10 000 from the RHS of the equation. Thus the equation will become:

$$10\,000 - 1000\,M - 10\,000 = 8000 - 10\,000$$

or, after simplification,

$$-1000\,M = 8000 - 10\,000. \qquad (4.6)$$

Comparing (4.5) with (4.6) we can see that the term 10 000 has 'moved' from the LHS to the RHS of the equation. In doing so its sign has changed from + to −. This can be generalised so that:

- **We can move a term from one side of the equation to the other so long as we change its sign in the process.**

(Many of you will no doubt recognise this as the rule you learned at school as 'change side, change sign'.)

Simplifying (4.6) we have:

$$-1000\,M = -2000. \qquad (4.7)$$

Notice now that if we divide the LHS of (4.7) by −1000 the coefficient of M will cancel out, leaving simply M. However, the division must be applied to both sides of the equation and so we have:

$$\frac{-1000\,M}{-1000} = \frac{-2000}{-1000}. \qquad (4.8)$$

You can see now why we decided earlier to leave zero out of the general rule about operations on equations. It would not be possible to divide both sides of an equation by zero. (We may multiply both sides of an equation by zero but the result will be $0 = 0$, which defeats the object of solving the equation!)

Now simplifying (4.8) we have $M = 2$. This is the root of the equation.

Algebraic manipulations of this kind can be followed to solve any linear equation involving one variable. We can generalise them as follows (assume the unknown variable in the equation to be solved is x):

1. Remove any fractions in the equation by multiplying both sides by a common denominator (such as the product of all denominators in the equation).
2. Using the rule 'change side, change sign' move all terms involving x to the LHS and all other terms to the RHS of the equation.
3. Combine terms on both sides of the equation, as far as possible.
4. Factorise the LHS of the equation using x as a common factor.
5. Divide both sides of the equation by the coefficient of x. This gives the root of the equation.

As you become more and more familiar with using steps 1. to 5. you may find yourself carrying them out in a different order, or ceasing to think consciously of them.

EXAMPLE

Solve:

$$2x/5 + 6x - 3/4 = -2x + 489/20.$$

SOLUTION

First remove the fractions by multiplying both sides of the equation by a common denominator. In this case 20 is the most efficient – we could use any multiple of 20, such as 40, but this would just result in unnecessarily large numbers. The equation now becomes:

$$8x + 120x - 15 = -40x + 489.$$

Collecting all terms involving x on the LHS, and all known terms on the RHS, we have:

$$8x + 120x + 40x = 15 + 489$$

or

$$168x = 504.$$

Now, dividing both sides of the equation by 168, we have:

$$x = 504/168 = 3.$$

So the root of the equation is 3.

It is good practice to check your arithmetic by verifying that the root you found does indeed satisfy the original equation. So here:

LHS = (2/5) × 3 + 6 × 3 − 3/4 = (24 + 360 − 15)/20 = 369/20.

RHS = −2 × 3 + 489/20 = (−120 + 489)/20 = 369/20.

Thus $x = 3$ makes the LHS and RHS of the original equation equal, and is therefore the correct solution.

Exercises 4.1

Solve the following equations:
(a) $3x + 1 = 2x + 3$;
(b) $2x + 3x/2 − 1 = 15x/2 + 5 + 2x$;
(c) $(3x + 15)/15 = (2x − 2)/12$;
(d) $x + 3 − 2x/3 = 4$.

4.3 Solving Quadratic Equations with a Single Variable

We saw in Chapter 3 that a quadratic function has the general form $y = ax^2 + bx + c$ where $a \neq 0$. Quadratic equations of a single variable have a very similar standard form which is:

$$ax^2 + bx + c = 0 \qquad (4.9)$$

where a, b and c are any numbers and $a \neq 0$.

A quadratic equation may already be in the standard form (4.9) for example, $x^2 − 2x + 3 = 0$ is in this form, with $a = 1$, $b = −2$ and $c = 3$. In other cases it will be necessary to perform some preliminary manipulations to express the equation in the standard form. For example:

$$10 = 3x^2 − 2x + 4,$$

is a quadratic equation, but to write it in the standard form we need to manipulate it into:

$$3x^2 - 2x - 6 = 0.$$

If after expressing a quadratic equation in the standard form (4.9) we find that $a = 0$ then the equation is no longer quadratic but linear, and can be solved by the methods of the last section. If either b or c or both are zero the equation remains quadratic, but it can be solved more simply than the general quadratic, as we shall see later. We will look first at methods for solving the general equation (4.9).

(a) The graphical method

EXAMPLE

Solve the following equation:

$$x^2 + x - 2 = 0. \qquad (4.10)$$

SOLUTION

We can solve this equation by graphical means in much the same way as we solved linear equations, by thinking of (4.10) as having been derived from the function:

$$y = x^2 + x - 2 \qquad (4.11)$$

to find the value(s) of x corresponding to $y = 0$. Thus to solve (4.10) we need to plot (4.11) and read off the value(s) of x corresponding to $y = 0$. The graph of (4.11) appears in Figure 4.3. We can see in Figure 4.3 that there are in fact *two* values of x which correspond to $y = 0$. They are $x = -2$ and $x = 1$, which are the points where the curve crosses the x-axis. Thus equation (4.10) has two roots.

It is possible for a quadratic function to have a graph which does not cross the x-axis. In such a case the corresponding quadratic equation has no (real) solution – that is, there are no values of x which satisfy the equation.

FIGURE 4.3
Graph of $y = x^2 + x - 2$

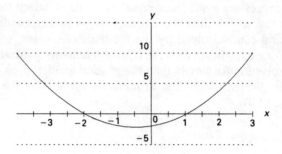

EXAMPLE

Solve the equation $x^2 + x + 2 = 0$.

SOLUTIONS

The solution to this equation would be given by the point(s) where the curve $y = x^2 + x + 2$ crosses the x-axis. The graph of $y = x^2 + x + 2$ appears in Figure 4.4. In fact, the curve never crosses the x-axis. That is to say, there are no values of x which will make the expression $x^2 + x + 2$ equal to zero. Thus the equation $x^2 + x + 2 = 0$ has no solution.

It is also possible for the graph of a quadratic function to be tangential to the x-axis. In such a case the corresponding equation has only one root.

EXAMPLE

Solve the equation $x^2 - 2x + 1 = 0$.

FIGURE 4.4
Graph of $y = x^2 + x + 2$

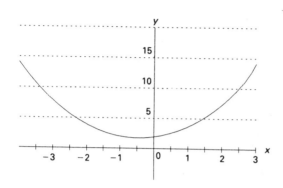

SOLUTION

The solution to this equation is given by the point(s) where the curve $y = x^2 - 2x + 1$ crosses the x-axis. The graph of $y = x^2 - 2x + 1$ appears in Figure 4.5. The curve is tangential to the x-axis at the point $x = 1$. (That is to say, the curve touches the x-axis at $x = 1$ but does not cross it.) Hence only $x = 1$ renders the expression $x^2 - 2x + 1$ equal to zero. Thus the equation $x^2 - 2x + 1 = 0$ has only one solution. (Another way of thinking of this case is to imagine a curve like that in Figure 4.3 moving upwards; the roots will get closer and closer together until eventually, when the curve is just tangential to the x-axis, they will coincide.)

In conclusion,

A quadratic equation may have zero, one or two real roots.

b) *The algebraic method*

As was the case with linear equations, we can use algebraic manipulations to solve quadratic equations more efficiently and more

FIGURE 4.5
Graph of $y = x^2 - 2x + 1$

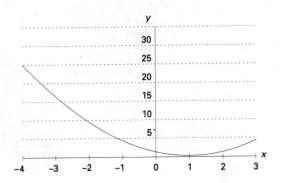

accurately. The simplest way to solve a quadratic equation, written in the standard form,

$$ax^2 + bx + c = 0$$

is to use the following formula;

$$x = (-b \pm \sqrt{b^2 - 4ac})/2a. \tag{4.12}$$

As noted earlier, a quadratic equation may not be expressed in the standard form to begin with. In this case it is necessary to manipulate it into the form (4.9) in order to be able to identify the values of a, b and c, before using formula (4.12).

For example, to solve the equation $x^2 - 2x = 3x + 2$ we need to manipulate it into the equivalent equation $x^2 - 5x - 2 = 0$. The latter has the form (4.9) with $a = 1$, $b = -5$ and $c = -2$. These values of a, b and c can now be used in the formula (4.12) to solve the equation.

Note that the formula involves the sign $[\pm]$. This is because, as we saw earlier, quadratic equations can have up to two roots. When there are two roots, one is obtained using the $[+]$ and the other the [−] sign. We will see below how the formula also applies in cases where there are fewer than two roots.

The reason why (4.12) gives the solution of the standard quadratic equation rests on the idea of factorising, which we met in Chapter 2. The detail of this process is beyond the scope of this book and it is not necessary to know it in order to use the formula.

EXAMPLE

Solve the equation $5/x^2 - 2 = 1/x + 4$.

SOLUTION

This equation is not expressed in the form $ax^2 + bx + c = 0$. It is therefore necessary to manipulate it into this form before we use formula (4.12) to solve the equation.

The fractions in the equation can be removed by multiplying both sides by x^2. The equation then becomes:

$$5 - 2x^2 = x + 4x^2.$$

Now transferring all terms to the LHS of the equation we have:

$$-6x^2 - x + 5 = 0.$$

We could leave the equation like this, but it is slightly easier to deal with if we multiply throughout by -1 to get:

$$6x^2 + x - 5 = 0.$$

This equation has the form:

$$ax^2 + bx + c = 0$$

with $a = 6$, $b = 1$ and $c = -5$.

Using these values in the formula (4.12) we find:

$$x = (-b \pm \sqrt{[b^2 - 4ac]})/2a$$
$$= (-1 \pm \sqrt{(1 - 4 \times 6 \times [-5])})/12 = (-1 \pm 11)/12.$$

Thus, the two roots are:

$$x = (-1 + 11)/12 = 5/6$$

and

$$x = (-1 - 11)/12 = -1.$$

You can check that the roots we have just obtained both satisfy the original equation.

We saw earlier that a quadratic equation may have no solution. What happens if we try to use the formula (4.12) to solve such an equation is that we find ourselves trying to take the square root of a negative number, because the expression $b^2 - 4ac$ in (4.12) takes a negative value. As we know, such a square root does not exist in the ordinary number system, and so we conclude that the equation has no solution. If the quadratic equation was derived from a real problem then either it has no possible solution, or a mistake has been made, perhaps in deriving the equation or during the process of solving it.

The other exceptional case we encountered above was when the quadratic equation had only one root. This occurs because the expression $b^2 - 4ac$ in the formula (4.12) is equal to zero. For example, try to apply the formula to the equation $x^2 - 2x + 1 = 0$ which we solved graphically earlier. Formula (4.12) in this case gives $x = (2 \pm \sqrt{(-2)^2 - 4 \times 1 \times 1})/2 = 1$, which is the single root of the equation.

(c) Solving simpler quadratic equations

We do not always need to resort to formula (4.12) or to a full graph of the corresponding function in order to solve a quadratic equation. If, when the equation is put in the standard form, either b or c turns out to be zero, then the equation can be solved more simply.

Take the case where $b = 0$. The standard form then reduces to:

$$ax^2 + c = 0.$$

This can be solved easily by solving for x^2 and then taking the square root of both sides of the equation to get the values of x.

EXAMPLE

Solve $4x^2 - 64 = 0$.

SOLUTION

Separating known from unknown terms we have:

$$4x^2 = 64.$$

Now solving for x^2 gives:

$$x^2 = 16.$$

Taking the square root of both sides we have the solution:

$$x = \pm 4.$$

Note that the solution ± 4 means both $x = 4$ and $x = -4$ satisfy the equation, so this quadratic has two roots.

Another, simpler, quadratic arises when in the standard form $c = 0$, giving:

$$ax^2 + bx = 0.$$

This equation can be solved very easily by factorising. We can take x out as a common factor, to get:

$$x(ax + b) = 0.$$

Now it is easy to see that this product will be zero if either $x = 0$ or $(ax + b) = 0$. The value of x that makes $ax + b = 0$ is $x = -b/a$; hence the two roots of $ax^2 + bx = 0$ are $x = 0$ and $x = -b/a$.

EXAMPLE

Solve the equation $2x^2 - 2x = x^2$.

SOLUTION

First we simplify the equation by moving all terms to one side. The equation becomes:

$$2x^2 - 2x - x^2 = 0.$$

Combining like terms reduces the equation to:

$$x^2 - 2x = 0.$$

Taking x out as a common factor gives:

$$x(x - 2) = 0.$$

The product is zero when $x = 0$ or $x - 2 = 0$, and $x - 2$ is zero when $x = 2$.
 Hence the two roots of the equation are $x = 0$ and $x = 2$.

Exercises 4.2

1. Solve the following equations:
 (a) $5x + 4x^2 - 54 = -2x^2 + 5x$;
 (b) $3x^2 - 6x = -x^2$;
 (c) $x^2 - x = 10x + 4(x - 1)$;
 (d) $1/x^2 - 2 = 7$;
 (e) $2x^2 + x(3x - 2) = 4$.

4.4 Solving a Single Equation Involving More than One Variable

When a single equation involves more than one variable we cannot obtain numerical values for the variables. All we can do is to express the value of one variable in terms of the rest.

EXAMPLE 1

Solve for x the equation $2x + 8y + 12z = 40$.

SOLUTION

We cannot resort here to graphs to solve the equation (at least not two-dimensional graphs). The function corresponding to the equation we need to solve would involve four variables, so to plot its graph we would need four dimensions!

We can easily solve the equation, however, using algebraic manipulations. To solve for x we shall treat all variables except x as if they are known. Moving 'known' terms to the RHS we have:

$$2x = 40 - 8y - 12z$$

and dividing both sides of the equation by 2 we get:

$$x = 20 - 4y - 6z. \qquad (4.13)$$

The interpretation of this solution is that any combination of values for x, y and z that satisfies the equation (4.13) is a solution of (i.e. also satisfies) the original equation.

We can assign arbitrary values to y and z and use (4.13) to get the corresponding value for x such that the values of x, y and z satisfy the original equation.

For example, we can arbitrarily set $y = 0$ and $z = 1$. Using (4.13) we get $x = 14$. Thus, the combination $x = 14$, $y = 0$ and $z = 1$ satisfies the original equation. This is just one of an infinite number of possible combinations satisfying the equation.

EXAMPLE 2

Solve the following equation for x in terms of a:

$$ax - 2 = x + a$$

SOLUTION

Here we are regarding x as the unknown and treating a as if it were known. Then separating known from unknown terms we have:

$$ax - x = a + 2.$$

Factorising the LHS we get:

$$x(a - 1) = a + 2.$$

Finally, dividing both sides by the coefficient of x gives:

$$x = (a + 2)/(a - 1).$$

This last equation gives the value of x in terms of a. For any given value of a the value of x can be computed: for $a = 2$ $x = 4$ and so on. The only exception to this is $a = 1$, for which x is undefined, since the denominator becomes zero.

Exercises 4.3

1. Solve for x the following equations:
   (a) $2x - 10 - 2a = 4ax$;
   (b) $3ax - 2 = 4x$;
   (c) $(a + 2)(x - 3) + 2 - a = 5x/2$.
2. Find two different combinations of values for x and y that satisfy the equation $x - 2y = 4(2x - y) - 2/5$.
3. Solve the equation $5a^2x^2 - 2ax = 0$ for x in terms of a. What values of a will ensure that the equation has two defined roots?
4. Solve for x the equation $y = 1/(1 - x)^2$.
5. A millionaire states in his will that his estate is to be shared between his wife, son and daughter so that his wife receives 75% of the total amount going to his daughter and his son. If the amount to be shared is £m million, express the son's share in terms of those of his mother and sister.

4.5 Simultaneous Equations

4.5.1 Introduction

So far we have concentrated on the solution of single equations with just one unknown variable. When we looked at a single equation with more than one unknown quantity, we found that any one of the unknowns could be expressed in terms of the others, but it was not actually possible to solve the equation in the sense of finding a numerical value for each of the variables.

The position is, however, very different if we have more than one equation as well as more than one variable. We call a set of equations involving several variables a set of *simultaneous equations*, because their solution is a set of values of the variables which satisfy all the equations at once – that is, simultaneously. Simultaneous equations may involve very large numbers of variables, and may be quadratic, cubic and so on in form. But we will concentrate mainly on the simplest case, involving just two linear equations with two unknowns – for example, the pair of equations:

$$x + 2y = 5$$
$$3x + 5y = 11.$$

If, as here, we have two equations involving two variables, then we will usually be able to solve the equations to obtain numerical values for the variables – though, as we will see, there are exceptions to this rule. More generally, we need n equations if we wish to solve for n unknowns. (Our earlier study of equations in one variable fits into this framework – one equation in one unknown can normally be solved.)

However, knowing that a solution to the pair of equations exists, and actually finding that solution, are two very different things. What we need is a systematic method for solving the equations. We will follow the pattern established in earlier sections, first examining the graphical approach to solving the equations, and then looking at algebraic methods.

4.5.2 Solving two linear equations in two unknowns

(a) The graphical method

In Chapter 3, we met the example of linear supply and demand functions, and found that the stable price, where supply = demand, corresponds to the point where the graphs of the two functions intersect. The stable price satisfied the supply equation (3.4) and the demand equation (3.6) simultaneously in that it made $D = S$.

The intersection of the graphs of functions can, in fact, be used

FIGURE 4.6
Graph of simultaneous equations

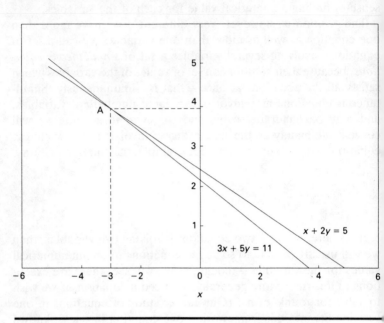

generally to solve simultaneous equations. Figure 4.6 shows the graphs of the pair of equations:

$$x + 2y = 5$$
$$3x + 5y = 11.$$

The easiest way to plot these equations is to use the points where x and y are zero; for example, in the first equation, when $x = 0$, $y = 2.5$, and when $y = 0$, $x = 5$, so the line can be drawn by joining the points $(0, 2.5)$ and $(5, 0)$. Remember from our definition of a graph in Chapter 3 that the graph of an equation is the set of all points satisfying that equation. It therefore follows that the point A with coordinates $(-3, 4)$ in Figure 4.6, lying as it does on the graphs of the equations, must satisfy both equations – that is, it represents the simultaneous solution of the pair of equations. You

FIGURE 4.7
Inconsistent simultaneous equations

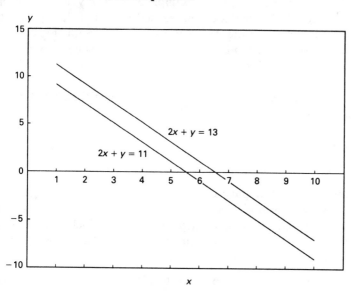

can verify by arithmetic that the values $x = -3$, $y = 4$ do indeed constitute a solution to both equations.

We can therefore say that:

- **The solution of a pair of simultaneous linear equations in two variables is given by the point of intersection of their graphs.**

The only circumstance in which two straight-line graphs will *not* intersect is when they are parallel; in this case the corresponding equations do not have a solution, and are said to be *inconsistent*. An example, involving the equations

$$2x + y = 11$$
$$4x + 2y = 26,$$

is shown in Figure 4.7. We will see why this situation arises when we look below at algebraic methods for solving the equations.

(b) The algebraic method

Since there are two algebraic methods available, both of which are widely taught, we will give an example of each. If you have a clear recollection of the method you learned at school, it is probably safest to continue using it; if you are coming to this topic 'from scratch', we recommend the first method as being perhaps slightly easier to get to grips with.

(i) The elimination method   Let us re-examine the pair of equations which we solved graphically:

$$x + 2y = 5$$
$$3x + 5y = 11.$$

If by algebraic manipulation we could combine the two equations in such a way as to get rid of one of the variables, we would arrive at a single equation in a single unknown, which our existing methods would then be adequate to solve. The following procedure will do the trick.

1. Multiply or divide one or both of the equations by suitable factors, in order to get the same coefficient of x in both equations. (The *coefficient* of x is, you will recall, the number in front of x.)
2. Add or subtract the equations in order to eliminate x and arrive at a single equation involving only y.
3. Use the methods of section 4.2 to solve this equation for y.
4. Substitute the value of y you have found into one of the original equations to find x.

This method sounds complicated when described in the abstract, but will be seen to be quite simple when we apply it to our sample equations. If you look back at them, you will see that as they stand the coefficients of x in the two are different. However, if we multiply the first equation by 3, while doing nothing to the second, then the coefficient of x in both of the new pair of equations will be 3:

$$x + 2y = 5 \text{ becomes } 3x + 6y = 15 \quad \text{(i)}$$
$$3x + 5y = 11 \text{ remains } 3x + 5y = 11 \quad \text{(ii)}.$$

Remember that in carrying out the multiplication you must multiply *all* the terms in the equation by 3 – including the RHS.

Having done this, we see that subtracting equation (ii) from equation (i) will eliminate x altogether, since $3x - 3x = 0$. That is exactly why we needed to have the same coefficient for x in both equations. The rest of the subtraction leads to:

$$(3x - 3x) + (6y - 5y) = 15 - 11,$$

so that $y = 4$. When you get used to this process, you probably will not need to write it down in so much detail.

The last step is to find the value of x by substituting $y = 4$ into one of the two original equations. It does not really matter which, but we choose the first because, having a coefficient of 1 for x, it gives us the value of x in the most direct way. The equation becomes:

$$x + 2y = x + 8 = 5, \text{ and so } x = 5 - 8 = -3.$$

The complete solution, then, is $x = -3$, $y = 4$. It is good practice to verify your solution by making sure that the values of x and y do indeed satisfy both the original equations. Here $x + 2y = (-3) + 2 \times 4 = 5$, and $3x + 5y = 3 \times (-3) + 5 \times 4 = -9 + 20 = 11$, as required. The solution also agrees with what we found graphically earlier in this section.

We have described the method, and illustrated it, in terms of eliminating x first, but of course there is absolutely no reason why you should not eliminate y instead if the coefficients make that the easier alternative. To see this, let us take a second pair of equations:

$$4x + 5y = 42 \quad \text{(i)}$$
$$3x - 4y = 16 \quad \text{(ii)}.$$

Here we have a case where there is not much to choose between the two variables in terms of ease of elimination. So for the sake of illustration we will eliminate y this time. We can do this by multiplying equation (i) by 4 and equation (ii) by 5, to get $20y$ in both:

$$16x + 20y = 168$$
$$15x - 20y = 80.$$

Since the coefficient of y is negative in the second equation, addition of the two equations will eliminate y:

$$(16x + 15x) + (20y - 20y) = 168 + 80$$

or

$$31x = 248,$$

whence $x = 8$.

Substitution in the original equation (i) then gives:

$$4 \times 8 + 5y = 42,$$

which we can solve to get $y = 2$. So the full solution is $x = 8, y = 2$.

(ii) The substitution method Recall our first pair of simultaneous equations:

$$x + 2y = 5 \quad \text{(i)}$$
$$3x + 5y = 11 \quad \text{(ii)}.$$

The alternative procedure for solving these can be summarised thus:

1. Solve the first equation for x in terms of y.
2. Substitute the resulting expression for x into the second equation, to get an equation involving y only.
3. Solve this equation for y by the usual methods for single-variable equations.
4. Substitute the value of y into the expression for x obtained in step 1. above to get the numerical value of x.

Step 1. in this case gives $x = 5 - 2y$. On substituting this expression for x into equation (ii), we find $3(5 - 2y) + 5y = 11$, so that $15 - 6y + 5y = 11$, or $15 - 11 = 6y - 5y$, giving $y = 4$, which agrees with what we found by the first method. Then $x = 5 - 2y = 5 - 8 = -3$ as before.

No matter which of the two methods you decide to use, you will nearly always end up with a unique solution – a single numerical value for each of x and y. However, there are a few exceptional cases which we need to think about. First, consider the equations:

$$2x + y = 11$$
$$4x + 2y = 26.$$

Dividing the second equation by 2 tells us that $2x + y = 13$. But this contradicts the first, which says that they total 11. Both of these equations cannot be true simultaneously, and so we say that the equations are *inconsistent* – there are no values of x and y which satisfy them both.

Even if you failed to notice this inconsistency, you should soon realise that something odd is going on. If you take the equations:

$$2x + y = 11$$
$$2x + y = 13$$

and subtract the first from the second, you obtain $0 = 2$ – not a very sensible result! This inconsistent case corresponds to the situation in Figure 4.7 where the graphs of the two equations are parallel, so that there is no point of intersection – and therefore no simultaneous solution.

Now consider a rather different case, typified by the equations:

$$x + 3y = 4 \quad \text{(i)}$$
$$3x + 9y = 12 \quad \text{(ii)}.$$

The second equation is simply the first multiplied by 3, so we do not really have two independent equations at all – just a single equation with two unknowns. As we saw earlier, the best we can do in this case is to express x in terms of y (or the other way round):

$$x = 4 - 3y \quad \text{(iii)}.$$

Then we can substitute any number we like for y, and find the corresponding value of x from equation (iii). For instance,

$$x = 4, y = 0; x = 1, y = 1; x = -26, y = 10$$

all satisfy the equations, as do an infinite number of other pairs of values. Simultaneous equations of this type are called *degenerate*.

Again, if you fail to spot degeneracy, it will be brought to your attention when you try to carry on with the solution of the equations. Suppose you decide to make the coefficients of x in the two equations above equal, by dividing the second by 3. Immediately you obtain a second equation identical with the first – so the method fails. The graphical interpretation of this case is that the lines representing the two equations have coalesced, so that instead of a single intersection representing a unique solution – there is a whole infinite set of solutions.

4.5.3 *More complex cases of simultaneous equations*

As we mentioned earlier, in general if we have as many equations as there are unknowns, we can find a unique solution (though there are higher-dimensional analogues of the exceptional cases mentioned above). In practice it is unlikely that you would encounter anything more complex than the two-variable case – at least, not without the aid of a computer package.

The situation sometimes arises where there are more equations than unknowns, in which case some of the equations may be superfluous to requirement, or *redundant*. On the other hand, there may be more unknowns than equations – a situation we have already encountered, giving rise to an infinite number of solutions obtained by expressing some variables in terms of others. We will not go into these cases in any detail here.

Finally, there is no reason why simultaneous equations must be linear. It is generally not too hard to solve a pair of equations in two variables in which one is linear and one quadratic; you will find an example in the exercises below, and if you think of the situation graphically you should see why such a pair of equations would give rise to not one, but *two* pairs of values satisfying the equations. Simultaneous quadratics, and equations with higher powers, can be messy or even impossible to solve, and you are not likely to come across them in practical applications.

Exercises 4.4

1. Solve the following pairs of simultaneous equations:
   (a) $x - 3y = 10$ (b) $3x + 5y = 15$ (c) $2s - 3t = 2$
       $2x + 5y = 42;$ $x + y = 3;$ $3s + 2t = 16;$
   (d) $2x + 5y = 13$ (e) $3a + 4b = 1$ (f) $x + y = 4$
       $4x + 10y = 15;$ $8a - b = 1;$ $2x + 2y = 8.$
2. Find a pair of numbers whose sum is 7, and whose difference is 15.
3. Two single and one return bus ticket for a certain journey cost £3.30, while two returns and three singles cost £5.70. How much does each type of ticket cost?
4. The weekly cost £C of producing N units of a product is given by $C = 500 + 3N$, and the weekly revenue £R generated by the sale of these units by $R = 6N(24 - N)$. Find (a) graphically; and (b) algebraically, the points at which revenue and costs are equal. What is the practical significance of these points?

4.6 Solving Linear Inequalities Involving a Single Variable

4.6.1 Introduction

In Chapter 1 we met the inequality signs, $(<, >, \geqslant, \leqslant)$, which are used for comparison of arithmetic quantities. For example, $3 < 6$ expresses the fact that 3 is less than 6, and so on.

Inequalities can also involve algebraic expressions, in which case they are closely related to equations, and can be handled in a similar way. An example will illustrate the similarity.

Consider the inequality $3x - 2 \leqslant 5 - x$. This allows $3x - 2$ to be either equal to or strictly less than $5 - x$. We can see that the inequality places less stringent requirements on the value of x than does the equation $3x - 2 = 5 - x$. This is because in the inequality *any* value of x which makes $3x - 2$ less than $5 - x$ is acceptable as a solution, not just the single one which makes $3x - 2$ equal to $5 - x$. The consequence of this is that (as we shall see below) in general the solution of an inequality consists of a range of values of the variable.

Inequalities arise in connection with many practical business problems. As with equations, they may involve more than one variable, and can be nonlinear, though in practical applications linear inequalities are more common. In this book we will only consider linear inequalities involving one variable.

4.6.2 *A method for solving linear inequalities with one variable*

By *solving* an inequality, we mean finding the value(s) of the unknown variable which satisfy the inequality – that is, which make the relation between the two sides of the inequality true. This is, of course, exactly the same definition as we had for the solution of an equation, so it is no surprise to learn that we can use our rules for solving linear equations to solve inequalities also – with the following important exception:

- **When both sides of an inequality are multiplied or divided by a negative number the direction of the inequality is reversed.**

It is easy to see that this rule makes sense. For example, if we multiply the two sides of $5 > 3$ by -1 they will become -5 and -3. Clearly $-5 < -3$, and the direction of the inequality is now reversed.

The remaining rules – about transferring a term from the left to the right side of an inequality, and about multiplying or dividing the two sides by a positive number – apply in exactly the same way to inequalities as they did to equations.

EXAMPLE 1

A flower shop has fixed daily expenses (staff, rent, heating, etc.) which amount to £100. Each customer spends on average £4.00. How many customers does the shop need each day if it is not to make a loss on that day?

SOLUTION

Suppose the number of customers needed each day is denoted by C. Then the revenue would be £$4C$ per day (assuming that each customer spends the average amount). If the shop is not to make a loss then we need:

$4C \geq 100.$

Dividing both sides of the inequality by 4 we get:

$C \geq 25.$

This is the solution of the inequality. It means that the number of customers needed each day is at least 25. (If the shop gets exactly 25 customers per day then it just breaks even. If the number of customers exceeds 25, the shop makes a profit; if there are less than 25 customers it makes a loss.)

Notice how solving the inequality has yielded an infinite number of acceptable values for the unknown variable C: all values of 25 and more are acceptable (though, of course, only integer values have a sensible practical meaning in this case). Had this been an equation, for example, $4C = 100$, then the solution would have consisted only of a single value, $C = 25$.

EXAMPLE 2

Solve the following inequality:

$$\frac{3x}{-5} - 2 \geq x - 3.$$

SOLUTION

We can remove the fraction by multiplying both sides by -5. This gives $3x + 10 \leq -5x + 15$. (Note the reversal of the inequality as a result of multiplying both sides by a negative number.)

Moving unknown terms to the LHS and known terms to the RHS of the inequality we have:

$3x + 5x \leq 15 - 10$

or

$8x \leq 5.$

Dividing both sides by 8 we obtain:

$x \leq 5/8.$

That is, any value of x equal to or less than $5/8$ satisfies the inequality.

4.6.3 *Multiple inequalities*

An inequality is merely a restriction or constraint on the values the unknown variable can take. It is possible that several inequalities can apply simultaneously in a practical situation, each one putting additional restrictions on the permissible values of the unknown variable. The situation can be illustrated by returning to our flower shop example.

EXAMPLE

Suppose the sales assistants in the flower shop in example 1 above cannot serve more than 100 customers in any one day. How many customers must the shop serve each day if it is not to make a loss on that day?

SOLUTION

We have already seen that if the shop is not to make a loss the number of customers C must satisfy:

$$c \geq 25.$$

The sales assistants cannot cope with more than 100 customers a day and so we must have:

$$C \leq 100.$$

To meet both restrictions, that, is not make a loss *and* serve no more customers than the sales assistants can handle, C must lie in the range from 25 to 100 inclusive. This range is written:

$$25 \leq C \leq 100.$$

The range we just determined is the solution to the two simultaneous inequalities $4C \leq 100$ AND $C \leq 100$. We have solved each inequality in turn to ascertain the range of values for C which satisfy both inequalities.

This approach can be extended to deal with any number of inequalities involving an unknown variable. Each one yields a range of values for the unknown variable. It is then necessary to identify where, if at all, those ranges overlap. The overlap will be the solution to all the inequalities simultaneously. We cannot display this situation by means of a two-dimensional graph, since we only have a single variable, x; however, the number line which we met in Chapter 1 may be useful in helping us to identify where the solutions to the individual inequalities overlap.

EXAMPLE

Determine the values of x that satisfy the following simultaneous inequalities:

(i) $2x - 3 \leqslant 15$;
(ii) $-4x + 2 \leqslant 2x - 10$;
(iii) $x - 3 \geqslant 2$.

SOLUTION

Solving each inequality in turn we have:

(i) $2x \leqslant 18$ so $x \leqslant 9$.
(ii) $-6x \leqslant -12$ so $x \geqslant 2$ (note the change from $\leqslant$ to $\geqslant$).
(iii) $x \geqslant 5$.

To see where, if at all, the three ranges of values for x overlap, we shade each range in turn on the number line (see Figure 4.8).

The three ranges overlap over the range:

$$5 \leqslant x \leqslant 9,$$

so this is the range of x values which satisfies all three inequalities.

In general, the more the inequalities x must satisfy, the narrower the range of acceptable x values will be. This can be seen clearly in Figure 4.8, where the range of x values satisfying all three inequalities is much narrower than the range for each inequality taken separately.

FIGURE 4.8
Plotting simultaneous inequalities on the number line

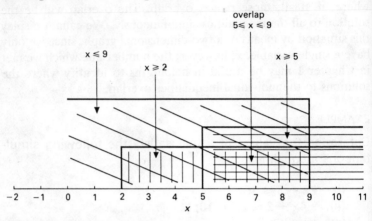

Exercises 4.5

1. Solve separately each of the following inequalities:
   (a) $x - 4 \geq 12$;   (b) $-3y - 4 \geq 3 - y$;
   (c) $2x - 5 \geq -\frac{1}{5} + x$; (d) $2x - 2 \geq 3x$.
2. What range of values for x satisfies the following simultaneous inequalities:
   (a) $3x - 4 \geq 2$;
   (b) $-4x - 6 \geq -14$; and
   (c) $4x - 2 \geq 1$?
3. What range of values for x in terms of a satisfies the following simultaneous inequalities:
   (a) $3(x - a) - 4 \geq a$;
   (b) $-4x + 8 \geq -12 - 20a$; and
   (c) $4x - a \geq 1$?
4. A farmer cultivates wheat on a 20-acre field. The yield of an acre fetches on average £500. The farmer can sell all the wheat he can harvest from the field but during the next season he can only devote 200 hours of his time and 30

hours of combine harvester time to the field. Each acre cultivated requires 12.5 hours of the farmer's time and 2 hours of combine harvester time. How much revenue can the farmer expect from the field next season?

5. A carpenter makes tables and chairs. He plans to work 40 hours next week. He is committed to making 5 tables during the week, each requiring 5 hours of his time. He can use the rest of his time to make chairs, each one requiring 3 hours. However, his supplies of timber will not permit him to make more than 4 chairs. What is the largest number of chairs the carpenter can make next week?

5 Introduction to Calculus

5.1 Introduction

We met the concept of the rate of change of a function in Chapter 3. We saw that the important feature of a linear function is that its rate of change with respect to the independent variable is constant. This means that linear functions have straight-line graphs with constant slopes which are easily calculated. We will see in this chapter that the rate of change of a non-linear function is not constant, and so its computation is more complicated. The computation of rates of change of nonlinear functions forms part of the topic known as *calculus*.

But first, let us spend a moment considering why we should be interested in calculating rates of change. A simple example will illustrate how important they are in practice. Recall the cost function (3.1) which we constructed in Chapter 3, giving production costs corresponding to different levels of car production at a plant. The rate of change of that function was constant at £6000, and represented the 'marginal cost of one car' – that is, the cost of producing one additional car beyond those already produced. In the same way, we can imagine the revenue function of the plant giving revenue levels corresponding to different levels of production. The rate of change of this function would give the 'marginal revenue from one car' – that is, the revenue from producing and selling one more car beyond those already produced and sold.

It is important for the firm to know the marginal revenue and marginal cost of a car. For example, suppose marginal revenue decreases with the level of production as the market becomes progressively oversupplied; then, in order to make a profit, the firm should set its production at a level where marginal revenue is greater than £6000, which was the marginal cost of producing a car. If production is increased beyond the point where marginal revenue is equal to £6000, the revenue from the manufacture and sale of one additional car will be lower than its cost of production, and so a loss will be made. If marginal cost is not constant at £6000 but varies with the level of production, the situation becomes more complicated but the basis of decision remains the same. (In fact, as we will see later, the firm can maximise its profits by fixing its level of production at the point where the marginal revenue from selling a car is exactly equal to the marginal cost of producing it.)

As with the cost and revenue functions at the car plant, so with many other functions: their rate of change is an important factor influencing operating decisions. Other examples are inventory costs as a function of the level of inventory, or procurement costs as a function of the amount procured.

Calculus, which forms the topic of this chapter, is concerned with the issue of rates of change and slopes in the general context of linear and nonlinear functions.

It is possible to define the rate of change of a function of several variables; however, our treatment of calculus will focus on functions of a single variable. Calculus has two main branches. One branch is known as *differentiation*, and deals with finding the rate of change of a function; this is covered in section 5.2. This process has many useful applications, for instance in finding local maximum or local minimum values of functions, as we will see in section 5.3. The second branch of calculus is known as *integration*. This is the inverse process to differentiation, and is used to find a function when we only know its rate of change. Integration falls largely outside the scope of this book. We will briefly introduce it in this chapter, and those interested in finding out more about integration and its uses can find further coverage in Appendix 1.

5.2 Differentiation

5.2.1 *Finding the slope of a curve*

As already mentioned, differentiation is a process which enables us to find the rate of change of a function.

We saw in Chapter 3 that in a linear function of one variable, written in general as $y = ax + b$, a is the rate of change of the function with x. We also saw that a is equal to the *slope* of the graph of the function $y = ax + b$. So the rate of change and the slope are, in fact, equal.

When we come to examine nonlinear functions, there is no corresponding general form of equation, and no equivalent to a for the rate of change. However, we can still define the rate of change of the function in the same way:

- **The rate of change of a function of one variable is equal to the slope of the graph of the function.**

There is just one further problem: a nonlinear function has a graph which is a curve, and it is not clear what the slope of the curve is!

We therefore need to begin by defining the slope of a curve. We do so in terms of the more familiar slope of a straight line, and say:

- **The slope of a curve at a given point is the slope of the straight line which is the tangent to the curve at that point.** (A tangent, in case you do not remember, is a line which touches the curve only at a single point.)

Figure 5.1 makes this definition clear.

So, in principle, to compute the slope of a nonlinear function we would have to draw the curve and then its tangent. Then we can compute the slope of the tangent as we did for straight line graphs of linear functions.

If you look again at Figure 5.1, you can see that at $x = 1$ the tangent is AB while at $x = 4$ the tangent is CD, and the lines AB and CD have different slopes. This exemplifies what we said in section 5.1: the slope of a curve varies along the curve. This does seem to leave us in a rather awkward position for determining the rate of change of a nonlinear function. We not only need to plot the curve, we also need to plot tangents to the curve at various points! However, *differentiation* provides a method for finding the slope at any point *without* any of this work.

We will begin by illustrating the process underlying differentiation, using the graph in Figure 5.2. Suppose we want to find the slope of this graph at point A where $x = 1$. This, as we saw above, is the slope of the tangent $A'B$ at A. We need to know two values of y along $A'B$ to compute its slope. We know the value of y at A

FIGURE 5.1
The slope changes along the curve

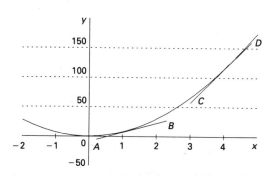

but we need one more y value. If we are to avoid plotting $A'B$ we can use a y value on the curve as an approximation for a y value on $A'B$. For example, we can use C as if it were a point on $A'B$. Then CD/AD would give us an approximation for the slope of $A'B$. Because both A and C are on the curve $y = x^2$, we can compute the values of y without plotting the curve and any tangent. Thus we see that the y-value at A is $1^2 = 1$, while at C it is $2^2 = 4$. Using our usual definition of slope of a straight line, we get:

approximate slope of $A'B = CD/AD = (4 - 1)/(2 - 1) = 3$.

Note, however, that CD/AD is the slope of AC and it is not a terribly good approximation to the slope of $A'B$, which is what we are after. But if we let C slide down the curve towards A, the slope of AC will get closer and closer to the slope of $A'B$. For example, from Figure 5.2, if C moves to C' which is at $x = 1.1$ the lines $A'B$ and AC' are indistinguishable and

slope of $AC' = AD'/C'D' = (1.1^2 - 1)/(1.1 - 1)$
$= 0.21/0.1 = 2.1$.

FIGURE 5.2
Computing the slope of $y = x^2$ at $x = 1$

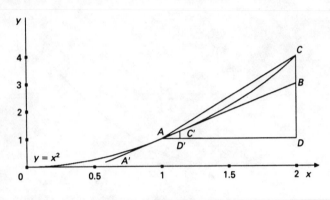

Try doing this for other positions of C – say, where $x = 1.01, 1.001$ and so on. You should find that the slope seems to be getting closer and closer to 2.

Now go through the same kind of argument to find the slope at $x = 2$. This time you should obtain a slope of 4. So it seems (though we are very far from having proved it mathematically!) that the slope of $y = x^2$ at any value of x is $2x$.

By going through this sort of process, but using symbols rather than specific numbers, we can arrive at a formula for finding the slope of the graph of any power of x:

- **The slope of the graph of $y = x^n$ at any value of x is given by nx^{n-1}.**

For instance, the slope of the graph $y = x^4$ is $4x^3$ (use the general formula with $n = 4$). Thus if we want to find the slope of this graph at $x = 2$, we evaluate $4x^3$ with $x = 2$ to get a slope of 32.

The slope of $y = x$ can be obtained by putting $n = 1$ in the formula (because $x = x^1$) to get slope $= 1x^0 = 1$. In this case the slope is a constant and it does not vary with x as in the example above. This should come as no surprise, because, of course, $y = x$ is a linear function giving a straight-line graph.

One other case is of particular interest. To find the slope of the graph of $y = 1$, remember that $x^0 = 1$, so the function can be viewed as $y = x^0$. Applying the formula to this, we find that the slope is $0x^{0-1} = 0$ – so the graph has zero slope for all values of x. This is, of course, correct, since the graph of $y = 1$ is a horizontal straight line and y does not change with x.

The process of using the formula to find the slopes of graphs of functions, and hence finding the rates of change of such functions, is called *differentiation*. We say that:

- **Differentiating x^n with respect to x gives nx^{n-1}.**

The process of differentiating $y = x^n$ with respect to x gives us a new function which is generally written as $y' = nx^{n-1}$. This new function is called the *derivative* of $y = x^n$ with respect to x.

We need a symbol to represent the differentiation operation (just as we have a symbol for addition, multiplication, etc.). The operation 'differentiate y with respect to x' is written **dy/dx**, and read as 'dee y by dee x'. Note that dy/dx is to be regarded as a single entity – so the ds cannot be cancelled! The reason for this notation is that traditionally dx has been used to mean 'an infinitesimally small increase in x', and the same for y. So dy/dx simply represents in shorthand the process of taking smaller and smaller increases in x and y in order to find the slope, as we did above using Figure 5.2.

We sometimes also write $d(x^2)/dx$ to mean 'differentiate x^2 with respect to x'; similarly, $d(t^3)/dt$ means 'differentiate t^3 with respect to t', and so on. This last calculation, incidentally, would give the answer $3t^2$; there is nothing special about the use of x as the variable in the differentiation formula.

To sum all this up:

1. The slope of the tangent to a curve at any point gives the rate of change of the function at that point.
2. The rate of change of a function, and the slope of its graph, are equal to the derivative of the function.
3. The derivative of y with respect to x is denoted dy/dx. Thus, if $y = x^n$, then $dy/dx = nx^{n-1}$.

5.2.2 *Some rules for differentiation*

We can extend the use of the differentiation rule to enable us to find the derivatives of more complex expressions. The following results can be proved by methods similar to those outlined above.

(i) The derivative of the product of a constant and a function is the product of the constant and the derivative of the function. Expressed symbolically, this means that:
if $y = ax^n$, where a is a constant, then $dy/dx = a\, nx^{n-1}$.
For example, if $y = 4x^2$ then $dy/dx = 4 \times 2x = 8x$, and so on.

(ii) The derivative of the sum of two functions is the sum of their derivatives. That is, if $y = x^n + x^m$, then
$dy/dx = nx^{n-1} + mx^{m-1}$.
For example if $y = 2x^2 - 4x$ then $dy/dx = 4x - 4$.

(iii) The derivative of any constant is zero. That is, if $y = k$, then $dy/dx = 0$, for any constant value of k. This follows from rule i, together with the fact noted earlier that the derivative of $y = x^0$ is zero.
For example, $y = 5$ is the same as $y = 5x^0$, and so $dy/dx = 5 \times 0 \times x^{-1} = 0$.

(iv) If $y = e^x$ then $dy/dx = e^x$.

This makes the natural exponential function $y = e^x$ rather special; the value of the function at a point $x = x'$ gives us the rate of increase of the function at $x = x'$.

There are many more rules for differentiation, dealing with products and quotients of functions, exponential and logarithmic functions, and so on, but we will not go any further here. Once you are familiar with the basic concepts you can always look up the formulae in more advanced books on calculus should the need arise. In fact, the rules given here will cover a great many practical situations.

EXAMPLES

(a) $d(6u^2 - 8u + 10)/du = 12u - 8$.
(b) $d(6/x + 2x - 7x^2)/dx = d(6x^{-1} + 2x - 7x^2) =$
$-6x^{-2} + 2 - 14x = 2 - 14x - 6/x^2$.
(c) $d((2 - t)^2)/dt = d((2 - t)(2 - t))/dt = d(t^2 - 4t + 4)/dt = 2t - 4$.
(d) What is the slope of the graph of the function $y = 2x^2 + 5x$ at $x = -1$?

The slope will equal the value of the derivative of the function $y = 2x^2 + 5x$ at $x = -1$. The derivative of $y(x)$ is $dy/dx = d(2x^2 + 5x)/dx = 4x + 5$. Thus the slope of the graph of $y = 2x^2 + 5x$ at $x = -1$ is $4 \times (-1) + 5 = 1$.

Exercises 5.1

1. Find dy/dx when y is equal to:
   (a) x^6;
   (b) $1/x^3$;
   (c) $\sqrt{(x^3)}$.
   (*Hint*: Express the functions in the form x^n before applying the formula.)
2. Find dy/dx when y is equal to:
   (a) $x^2 + 2\sqrt{x}$; (b) $3x(x - 2) + x^{3/2}$.
3. Differentiate $y = (x - 2)(x - 3)$ with respect to x.
4. Find dv/dt if $v = 6 + 8t$.
5. What is the slope of $y = 75x - x^3$ at the point where $x = 2$?
6. Find dy/dx when $y = x^2 - 4x + 2$. Then find the value of x when $dy/dx = 0$. What does this tell you about the rate of change of $y = x^2 - 4x + 2$?
7. A farmer has a strawberry crop of 1000 tonnes. She estimates that each day she stores the crop 10% of the initial weight is lost due to rot. However, the price at which she can sell the crop rises by £100 per tonne per day stored. The initial price is £2500 per tonne.
   (a) Express the revenue £R the farmer can get from the crop as a function of the number of days she stores her crop.
   (b) What is the rate at which this revenue is rising with the number of days the crop is stored? Comment on your result.

5.2.3 Higher order derivatives

It is clear from what we have seen of differentiation so far that the derivative of a function is often a function itself. We may well therefore be interested to know the rate of change of the derivative. Differentiating the derivative will give its rate of change. For example, if we have a function giving the distance travelled by a car as a function of the time it has been travelling, then its derivative gives the rate of change of that distance with time, better known as the speed of the car. Differentiation of this derivative with respect to time gives the rate of change of the speed of the car with time, better known as the acceleration of the car.

The derivatives resulting from repeated differentiation of a function are known as its *higher order* derivatives. Upon first differentiation of a function we obtain the *first derivative* of the function. For example, if $y = 3x^3$ its first derivative is $dy/dx = 9x^2$.

If we now differentiate dy/dx itself with respect to x, the derivative obtained is called the *second derivative* of the function $y = 3x^3$, and is denoted d^2y/dx^2. This is read 'dee 2 y by dee x squared'. Thus we have:

$$d^2y/dx^2 = d(9x^2)/dx = 18x.$$

The interpretation of the second derivative is similar to that of the first: it gives the rate of change of the first derivative. The second derivative will come in useful in finding where a function has a local maximum or minimum, as we will see in the next section. This process of repeated differentiation can be continued to give any order derivative we wish.

A function may not have a derivative at every point. For example $y = 1/x$ is discontinuous (not defined) for $x = 0$, as we saw in Chapter 3. Its derivative $dy/dx = -1/x^2$ is also undefined at this value of x, as indeed are all its higher order derivatives.

Exercises 5.2

1. Find the second derivatives of the functions in question 2 of Exercises 5.1.
2. Find d^2y/dx^2 and d^3y/dx^3 when:
    (a) $y = x^3 - 2x + 3$;
    (b) $y = x^{-2}/2 + 4x$;
    (c) $y = 3x^2$.
3. A car sets off at time zero, and spot checks on the distance it covers from its starting point reveal that when it has travelled for t hours the distance s (km) is given by the expression $s = 60t + 6t^2$.

(a) What is the speed of the car at time t?
(b) What is the rate of acceleration/deceleration of the car half an hour after it sets off?

5.3 Optimising a Function

One of the most important uses of differentiation is in helping us find where a function has a minimum or a maximum value. We refer to this process as *optimising* the function. Optimisation is very important in practical situations; for example, if profit is given by a non-linear function of the number of units produced, then it is important to know the level of production which will maximise profits. This section explains how derivatives can be used to optimise functions.

EXAMPLE

Refer to the publisher in Question 2 of Exercises 3.2. You should have deduced there that when the publisher charges $(0.65 + P)$ pounds a copy the revenue is
$R = (0.65 + P)(500 - 600P)$.

What price should the publisher charge to maximise revenue?

SOLUTION

We will solve this question graphically, and then see how derivatives could have been used to avoid the need to draw the graph. Figure 5.3 shows revenue as P ranges from 0 to 0.8. Clearly, when P is about 0.09 (and hence the price charged is about £0.74), revenue is maximised at £330 000 per day. However, two questions arise: (i) can we identify the revenue-maximising price more accurately; and (ii) can we do so without the use of the graph?

The answer to both questions is 'yes' if we remember that the derivative of the function R above with respect to P will give us the rate at which revenue changes as P rises. We see

FIGURE 5.3
Optimising R = (0.65 + P)(500 − 600P)

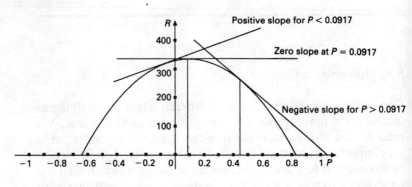

next how we can use the derivative of R to find where R takes a local maximum value.

Differentiating R with respect to P and simplifying, we obtain:

$$dR/dP = 110 - 1200P. \qquad (5.1)$$

If dR/dP is positive, revenue rises with increasing P, and if dR/dP is negative, revenue falls as P increases. (This is the same interpretation as we had for slopes or rates of change of linear functions in Chapter 3).

Clearly, when $P = 0$, $dR/dP = 110$, which means that initially revenue rises with P. As P increases, however, dR/dP will decrease (because P has a coefficient of −1200 in (5.1)), and will eventually reach zero. We can find the value of P at which this occurs by solving the equation:

$$dR/dP = 110 - 1200P = 0.$$

This gives $P = 0.0917$.

As P rises above 0.0917, dR/dP becomes negative. This means that now revenue falls as P rises, because the higher price per copy does not compensate for the lost sales.

What we have shown, then, is that the derivative starts off large and positive at $P = 0$, then decreases to zero at $P = 0.0917$, and becomes negative if P is increased further. You can follow this on the graph in Figure 5.3. The slope of the curve is upward (positive) and quite steep at $P = 0$; the curve becomes flatter as we move to the right, reaches a slope of zero at $P = 0.0917$ (where the tangent to the curve is parallel to the P-axis), and then begins to go downwards (negative).

Thus so long as the initial price of £0.65 (65 pence) is increased by anything up to 0.0917 (approximately 9 pence), revenue continues to rise. If the price rises by more than 9 pence, revenue begins to fall. *So revenue will be a maximum when the price has risen by 9 pence, to the nearest penny.* Thus the revenue-maximising price will be 65 + 9 pence (74 pence).

We have now identified the value of P which maximises revenue without needing to use the graph. This process can be generalised to enable us to identify local maximum or minimum values of any function.

The key feature at a local maximum point, as we have just seen, is the following:

- **The first derivative of the function is positive just to the left of the local maximum point, zero at the local maximum point, and negative just to the right of the local maximum point.**

This is illustrated in Figure 5.4, on which the function $R = (0.65 + P)(500 - 600P)$ and its derivative $dR/dP = 110 - 1200P$ are both plotted. The derivative is, of course, a straight line, and as you can see it does indeed cross the P-axis from above to below precisely at the maximum point of the graph of R.

These observations give us a procedure for identifying a local maximum:

1. Differentiate the function.
2. Set the derivative equal to zero.
3. Solve the resulting equation to determine the value of the independent variable.
4. Test either side of that value to check whether the derivative changes from positive to negative as we move from left to right.

FIGURE 5.4
Graph of R = (0.65 + P)(500 − 600P) and dR/dP = 110 − 1200P

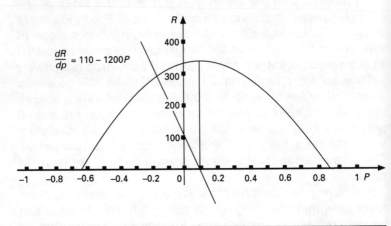

We can, however, be much more elegant and obviate the need for step 4. Notice in Figure 5.4 that the first derivative dR/dP is *decreasing* with P at the local maximum point. Therefore *its own* derivative, which is the *second* derivative d^2R/dP^2 of the original function R, must be *negative* at the local maximum point.

Generalising this sequence of thoughts, we arrive at the procedure which is normally used to identify a local maximum point of any function $y = f(x)$. This can be summarised as follows:

1. Find dy/dx.
2. Solve $dy/dx = 0$ and let x' be a root of this equation.
3. Find the value of d^2y/dx^2 at $x = x'$.
4. If d^2y/dx^2 at $x = x'$ is negative, then $y = f(x)$ attains a local maximum value at $x = x'$.

In the publisher's case the second derivative is $d^2R/dP^2 = -1200$. This is a negative constant, and so d^2R/dP^2 is always negative; thus it will certainly be negative at $P = 0.0917$. The conditions for a local maximum point at $P = 0.0917$ are therefore met.

In exactly the same way, we can deal with the case of a local minimum, such as that illustrated in Figure 5.5. You can see that to the left of the local minimum point, the function is decreasing,

FIGURE 5.5
The local minimum of $y = 3x^2 - 12x$ is at $x = 2$

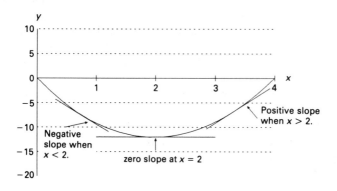

and so its derivative will be negative. To the right of the local minimum point, the derivative will become positive in line with the increasing function. Clearly, at the local minimum point where the function changes from a decreasing to an increasing one, the derivative will be zero. Thus in the same way as we derived the procedure for finding a local maximum, we have the following method for finding a local minimum of the function $y = f(x)$:

1. Find dy/dx.
2. Solve $dy/dx = 0$ and let x' be a root of this equation.
3. Find the value of d^2y/dx^2 at $x = x'$.
4. If d^2y/dx^2 at $x = x'$ is positive, then $y = f(x)$ attains a local minimum value at $x = x'$.

Figure 5.5 shows the local minimum point of the function $y = 3x^2 - 12x$. The first derivative $dy/dx = 6x - 12$ is zero for $x = 2$. The second derivative is $d^2y/dx^2 = 6$, which is positive for all values of x, including $x = 2$. Thus at $x = 2$ we have a local minimum point.

When using the first and second derivatives to test for a local maximum or minimum point, you may find that the second derivative at that point turns out to be zero. In such a case it is advisable to evaluate the first derivative on either side of that point. If the first derivative changes from positive to negative we have a local maximum point; if it changes from negative to positive we have a local minimum point. *If the first derivative does not change sign, then we have neither a local maximum nor a local minimum.* Such points, where the first derivative is zero but does not change sign, are known as *points of inflexion*. The key feature at a local maximum or minimum is the fact that the function 'turns round', from decreasing to increasing or vice versa. The local maxima and minima are therefore collectively known as *turning points* of the function.

One more example will help to demonstrate how we identify local maximum and minimum points.

EXAMPLE

Find the turning points of $y = x^3/3 - x^2/2 - 6x + 5$ and state in each case whether the point is a local maximum, minimum or a point of inflexion.

SOLUTION

Following the four-step process outlined earlier we proceed as follows:

1. *Find dy/dx*.
   We have $d(x^3/3 - x^2/2 - 6x + 5)/dx = x^2 - x - 6$.
2. *Solve dy/dx = 0 to find the root(s) of the equation*.
   Solving $x^2 - x - 6 = 0$ by using the formula (4.12) for solving quadratic equations we find two roots: $x = 3$ and $x = -2$.
3. and 4. *Find the value of d^2y/dx^2 at the roots of dy/dx and hence determine the nature of each turning point*.
   We have $d^2y/dx^2 = d(x^2 - x - 6)/dx = 2x - 1$.
   Clearly at $x = 3$, $d^2y/dx^2 = 2 \times 3 - 1 = 5$. Since $5 > 0$ the function $y = x^3/3 - x^2/2 - 6x + 5$ has a local *minimum* at $x = 3$.
   At $x = -2$ we have $d^2y/dx^2 = 2 \times (-2) - 1 = -5$. Since $-5 < 0$ the function $y = x^3/3 - x^2/2 - 6x + 5$ has a local *maximum* at $x = -2$.

FIGURE 5.6
Local and absolute optimal points

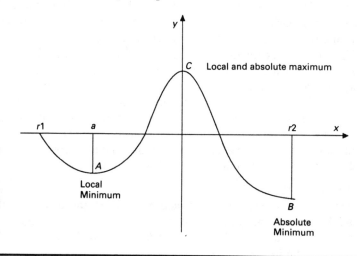

If you care to plot the function $y = x^3/3 - x^2/2 - 6x + 5$ you should be able to verify the above findings.

5.2.4 *Local and absolute maxima and minima*

We can now clarify why we have been referring to points as a *local* maximum or minimum rather than simply a maximum or minimum. In colloquial language, when we say 'maximum' or 'minimum' we mean the largest or smallest values of a quantity. However, it cannot be assumed that because a function has a local minimum at a certain point, this will necessarily be *the* smallest value of that function. This is illustrated by Figure 5.6, which shows a function plotted over the range $x = r1$ to $x = r2$. Point A is certainly a local minimum as defined above, but equally certainly is not the smallest value of the function in this range – that distinction belongs to point B. Point B is what is known as the *absolute minimum* of the function in the range $r1$ to $r2$, in contrast with point A which is merely a local minimum turning point. In a similar manner we can see that point C is the *absolute maximum* of the function in the range $r1$ to $r2$, as well as being a *local maximum*.

In practical situations we are usually more interested in the absolute maximum or minimum value of a function, rather than in local maxima or minima. For instance, the publisher in the example given earlier will be interested in the price that maximises revenue within the full range of prices open to him.

The first and second derivative tests we discussed earlier identify only local maxima and minima. These, as we saw in Figure 5.6, may or may not coincide with the absolute maximum and minimum of the function in a given range. You can see by looking again at the figure that if the absolute maximum does not coincide with a local maximum, then it must occur at either $r1$ or $r2$. The same will be true of the absolute minimum. The following rule will help identify absolute maximum and minimum points:

- **To identify the absolute minimum (maximum) of a function in the range $r1$ to $r2$, evaluate the function at $r1$, $r2$ and at any local minimum (maximum) points in the range. The lowest (highest) value obtained will be the absolute minimum (maximum) value of the function in that range.**

EXAMPLE

A hairdresser has found that when she charges £5 per haircut she gets 100 clients a week. If she lowers the price to £4 she gets 120 clients a week. Assume that the number of clients per week is a linear function of the price £P.

(a) Write the revenue as a function of price.
(b) What is the marginal revenue when the price is £4.50, £5.50?
(c) The hairdresser will consider charging any price in the range £4 to £7. What price will minimise her revenue? What price will maximise her revenue?

SOLUTION

(a) The revenue is $R = C \times P$, where C denotes the number of clients per week, and P denotes the price in £s.

If clients are a linear function of price, the function can be written $C = aP + b$, where a is the rate of increase of the function with P. We are told that as P rises from £4 to

£5, 20 clients are lost. So $a = -20$. We are also told that when $P = 4$, $C = 120$. So $120 = -20 \times 4 + b$. Hence $b = 200$. So C as a function of P is

$$C = 200 - 20P.$$

The revenue function is:

$$R = C \times P = (200 - 20P) \times P = 200P - 20P^2.$$

b) The marginal revenue at a given price is the rate at which revenue will rise as the price is increased from its current level. Marginal revenue is therefore given by the derivative of R with respect to P.

Marginal Revenue $= dR/dP = d(200P - 20P^2)/dP$
$= 200 - 40P.$

So when $P = £4.50$ marginal revenue is $200 - 40 \times 4.50 = £20$.
When $P = £5.50$ marginal revenue is $200 - 40 \times 5.50 = -£20$.
This means that if the price rises by a small amount ε beyond £4.50, weekly revenue rises by £20 $\times \varepsilon$. Note that ε has to be marginal, i.e. very small, if £20 $\times \varepsilon$ is to be a reasonably accurate estimate of the increase in revenue. This is because the marginal revenue is itself a function of price, and so as soon as we change the price, the marginal revenue also changes! Thus we can fairly say that as price rises by 1 penny, weekly revenue rises by approximately 20 pence. However, if P rises by £1 it would be very inaccurate to claim that revenue will rise by £20. You can work out the revenues at £4.50, £4.51 and £5.50 to see for yourself how this decreasing accuracy works.

c) We need the absolute maximum and minimum value of the function R in the range £4 to £7.

To determine the absolute maximum, let us first investigate whether R has any local maximum in the range £4 to £7 – that is, if $dR/dP = 0$ and $d^2R/dP^2 < 0$ at any point in this range.

We have $dR/dP = 200 - 40P$ which is zero when $P = 5$. The second derivative of $R(P)$ is $d^2R/dP^2 = -40$. Thus the function $R(P)$ has only one turning point, at $P = 5$, which is a local maximum.

The absolute maximum in the range £4 to £7 will either be at $P = £5$ or at one end of the range. Evaluating R we have:

$R(P = £4) = £480.00$
$R(P = £5) = £500.00$
$R(P = £7) = £420.00$.

Thus in the range £4 to £7 the absolute maximum of R is at $P = £5$ and the absolute minimum at $P = £7$. Had R had a local minimum in the range £4 to £7 it would have been necessary to evaluate the function at that point too before the absolute minimum point could be identified.

Thus, in the price range £4–£7 per haircut the hairdresser will have minimum weekly revenue of £420 if she charges £7 per haircut, and maximum weekly revenue of £500 when she charges £5 per haircut.

EXAMPLE

Reconsider the function $y = x^3/3 - x^2/2 - 6x + 5$ of the previous section. Find its absolute maximum and minimum values in the range $-4 \leq x \leq 4$.

SOLUTION

Following the process outlined in the above section we proceed as follows:

(a) Absolute minimum value.
   This will occur either at one of the two ends of the range -4 to 4 or at the local minimum point within the range which we found was at $x = 3$. Evaluating the function at these three points we have:

   $y(-4) = -1/3$
   $y(3) = -17/2$
   $y(4) = -17/3$

So the absolute minimum of the function $y = x^3/3 - x^2/2 - 6x + 5$ in the range $-4 \leq x \leq 4$ occurs at $x = 3$, which in this case was also a local minimum point.

(b) Absolute maximum value.
This will occur either at one of the two ends of the range -4 to 4 or at the local maximum point within the range, which we found was at $x = -2$. Evaluating the function at these three points we have:

$y(-4) = -1/3$
$y(-2) = 37/3$
$y(4) = -17/3$

The absolute maximum of the function $y = x^3/3 - x^2/2 - 6x + 5$ in the range $-4 \leq x \leq 4$ therefore occurs at $x = -2$, which was also a local maximum point.

Exercises 5.3

1. Find the turning points of the following functions, stating in each case whether they are local maxima, minima or neither.
    (a) $y = 0.5x^2 + 2x + 20$.
    (b) $y = x^3 - 6x^2 + 12x - 7$.
    (c) $y = x^3 - 6x^2 - 5$.
2. A charter airline offers discounts to travel agents in order to attract business. If an agent books up to 50 seats to a destination the airline charges £300 per seat. If the agent books in excess of 50 seats there is a discount on all seats which is £1 per booked seat in excess of 50. There are only 300 seats on the plane.
    (a) How many seats should an agent book to minimise the cost per seat?
    (b) How many seats should the airline sell to an agent to maximise its revenue from that agent?
3. In Question 2 above, what is the marginal revenue to the airline from selling an additional seat to an agent who has already booked 80 seats? Would the airline be better off

offering 81 seats to this agent or the 81st seat to an independent traveller offering £100 for it?

4. The daily cost of producing t tonnes of paint at a factory is estimated to be given by the equation $C = t^3/3 - 6t^2 + 32t + 25$ where C is in £000s. Production is to be at least 0.1 tonnes and no more than 10 tonnes per day. Calculate the following:
   (a) The production level at which daily cost of production is a minimum.
   (b) The production level at which daily cost of production is a maximum.
   (c) The marginal cost of production when the factory produces 5 tonnes per day. Comment on your result.

5.4 Integration

5.4.1 Background

As we have seen, differentiation is the process of finding a derivative given a function. *Integration* is the inverse operation to differentiation, enabling us to find a function given its derivative. Integration is particularly useful when the rate of change of some quantity is known; for example, we may know how maintenance costs per time period vary with the age of a piece of machinery, and need to know total maintenance costs over the life of the machine.

Generally speaking, integration is a more complex process than differentiation – for example, we can always find the derivative of an algebraic expression, but there are many functions which cannot be integrated except by using a computer. In this section we give a brief discussion of integration, using functions of the form x^n as an illustration. Those of you who would like to know a little more about integration and its applications may wish to read Appendix 1, or to consult a text giving a fuller treatment of introductory calculus.

5.4.2 The process of integration

Integration means the process of finding a function given its derivative, and is denoted by the symbol $\int$. Thus $\int x^2 \, dx$ means 'find the function which when differentiated with respect to x has a derivative of x^2'. The process of finding the integral is known as *integrating x^2 with respect to x*. The expression $\int x^2 \, dx$ is called an *integral*, and the 'dx' in the integral represents 'with respect to x'. As in differentiation, so in integration dx is a single symbol and neither the d nor the x can be combined with anything else inside the $\int$ sign. As usual, there is nothing magic about the letter x; there is no reason why we should not find $\int y^2 \, dy$ or $\int n^2 \, dn$ instead.

We can deduce some rules for integration by simply reversing the rules for differentiation. Thus to find the integral of x^2 with respect to x, we note that if $y = x^3$ then $dy/dx = 3x^2$. This differs from the function we wish to integrate only by the coefficient 3. So making a suitable adjustment to the coefficient, we get:

$$\int x^2 \, dx = x^3/3$$

because $d(x^3/3)/dx = x^2$.

However, $x^3/3$ is not the only function which when differentiated gives x^2; we also get x^2 if we differentiate, for example, $x^3/3 + 10$. In fact, because the derivative of a constant with respect to x is always zero, we can say that *any* function of the form $x^3/3 + K$, where K is a constant, will have derivative x^2. So, in general,

$$\int x^2 \, dx = x^3/3 + K$$

where K is any constant. We call the expression $x^3/3 + K$ the *indefinite integral* of x^2 with respect to x, since it contains the arbitrary constant K and so is not completely defined.

Similarly, we can get a rule for integrating any power of x (except, as we will see, x^{-1}) by inverting the result $d(x^n)/dx = nx^{n-1}$:

$$\int x^n \, dx = x^{n+1}/(n+1) + K, \text{ for all values of } n \text{ except } -1.$$

K is called the *constant of integration*. This rule can be readily verified by noting that if $y = x^{n+1}/(n+1) + K$ then $dy/dx = x^n$. When $n = -1$, $n + 1 = 0$, which would make the integral undefined, so the rule does not hold for $n = -1$. It can be shown in fact that:

$$\int x^{-1} \, dx = \int 1/x \, dx = \ln x + K,$$

where $\ln x$ is the natural logarithm mentioned in Chapter 3. The proof of this result is, however, beyond our scope in this book.

Some very useful rules for integrating more complex expressions can be derived from the rules for differentiation obtained earlier:

1. $\int a \, dx = ax + K$, for any constant a.
2. $\int ax^n \, dx = a \int x^n \, dx = ax^{n+1}/(n+1) + K$, for all values of n except -1.
3. $\int (g(x) + f(x)) \, dx = \int g(x) \, dx + \int f(x) \, dx$.
4. $\int e^x \, dx = e^x + K$.

All these can be checked by differentiation.

Now let us look at a few examples of these rules in action.

EXAMPLES

(i) $\int 3x^2 \, dx = 3 \int x^2 \, dx = 3x^3/3 + K = x^3 + K$,
using rule 2. above.
You can verify that this is true by differentiating: when $y = x^3 + K$, $dy/dx = 3x^2$.

(ii) $\int (3x^2 + 4x^3) dx = 3 \int x^2 \, dx + 4 \int x^3 \, dx = x^3 + x^4 + K$.
This can readily be verified since $d(x^3 + x^4 + K)/dx = 3x^2 + 4x^3$.

Example (ii) illustrates rule 3. above, which states that the integral of the sum of two functions is the sum of their separate integrals. The rule can be extended to the sum or difference of any number of functions, but NOT to the product of functions. So we *cannot* say $\int f(x) g(x) \, dx = \int f(x) \, dx \int g(x) dx$. There are more complex rules for integrating functions of this kind, and other more complicated expressions, but the three we have mentioned will cover many of the practical cases you are likely to encounter.

5.4.3 Some uses of integration

As we have seen, integration enables us to determine a function from its derivative, which is the function giving its rate of change. Thus, in practical situations where the rate of change of a quantity is known, we can use integration to determine total quantity accumulated. Two simple examples will serve to illustrate the process.

EXAMPLE 1

Suppose a company producing printers for computers knows that the cost of producing one additional printer is £300, irrespective of how many printers have already been produced. This is equivalent to saying that the rate of increase of costs with production is 300. The company is interested in finding an expression for its total cost function.

If we denote total weekly cost in £ by C, and number of units produced per week by n, then the rate of increase of C with n can be written as dC/dn, and so $dC/dn = 300$. So C is the function which, when differentiated with respect to n gives 300. But we now know that to find such a function we need to integrate 300 with respect to n:

$$C = \int 300 \, dn = 300n + K, \qquad (5.2)$$

using rule 1. above. This, however, cannot be the whole answer; the indefiniteness introduced by the constant K makes equation (5.2) useless to the firm as a representation of total costs. We need some further information to enable us to determine a specific value of K. If, for example, the firm also knows that when no printers at all are produced, there are still fixed costs of £8000 per week to be met, then we can say that $C = 8000$ when $n = 0$, and substituting this into (5.2) leads us to conclude that in this case $K = 8000$. The expression for total costs is therefore $C = 300n + 8000$.

EXAMPLE 2

This final example will give you some feeling for what the constant of integration might mean in geometrical terms.

Find the equation of the function of x whose graph has a slope given by $x + 2$, for any value of x.

Suppose the equation we are looking for is denoted by $y = f(x)$. Then we know that the slope $dy/dx = x + 2$. So we can use integration to say that:

$$y = \int (2x + 2)\, dx = x^2 + 2x + K,$$

where K is the constant of integration (we have used rules 1. and 3. here). Thus $y = x^2 + 2x + K$ is the function whose graph has a slope of $2x + 2$. You should recognise it as being a quadratic function, since it contains only powers of x of which the highest is a square.

In fact this equation, due to the presence of the arbitrary constant K, actually represents not just one quadratic curve but a whole infinite family of curves, some of which are shown in Figure 5.7. They are all 'parallel', so that they have the same slope function, but the actual 'position' of the curve will be determined by the value given to K.

For example, suppose we want to find the function with slope $2x + 2$ which passes through the point (1, 1). We already know that it will be of the form $y = x^2 + 2x + K$, but now we also know that $y = 1$ when $x = 1$. So we must have $1 = 1 + 2 + K$, from which we see that $K = -2$. The required equation is therefore $y = x^2 + 2x - 2$. As in the previous example, the extra piece of information – in this case, knowing one specific point on the curve – has enabled us to eliminate the constant of integration.

The process involved in finding other types of integral, and applying them to the solution of practical problems, is briefly outlined in Appendix 1, but for a detailed treatment you would need to consult a text on elementary calculus. We hope, however, that the content of this section has been sufficient to give you at least the flavour of the topic.

Exercises 5.4

1. Determine the following integrals:
    (a) $\int 2x\, dx$;

FIGURE 5.7
A family of parallel curves

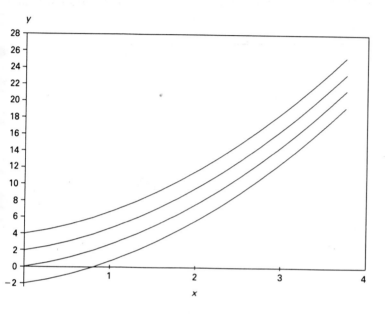

(b) $\int t^{-5}\, dt$;

(c) $\int (x^2 + 5x + 6)\, dx$;

(d) $\int 5/y\, dy$.

2. Determine the following integrals:
   (a) $\int 1/\sqrt{x^3}\, dx$;

   (b) $\int (z)^{1/2}\, z\, dz$;

   (c) $\int (x + 1)^2\, dx$.

3. A company knows that the cost of producing one more unit of its product depends on the current level of production. If n is the number of units already produced then the cost of the next unit is $C = 20 - 0.01n$, where C is in £s. The company has fixed costs of production of £1000. Determine the function giving the company's total costs of

production as a function of n, the number of units produced. (*Hint*: Note that C is the *marginal* cost of production. Recall our earlier discussion of marginal revenue.)
4. Find the equation of the curve which has slope $1/x^2$ and passes through the point where $x = 1$, $y = 0$.

6 Solving Practical Problems with Mathematics

6.1 Introduction

Many people who are beginning to use mathematics to help them solve practical problems experience the same difficulties. These difficulties are not so much with the mathematical techniques themselves, which may be quite straightforward when viewed simply as calculations. The harder part is, first, the process of translating the practical problem into mathematical form, and second, the interpretation of the mathematical solution back into useful practical information. If we think of the process of using mathematics on a 'real-world' problem in pictorial terms as shown in Figure 6.1, then it is the transitions from the real to the mathematical world and back which are the hard part.

If we look at the process represented in Figure 6.1 in a bit more detail, we see that there are three major steps involved:

1. Translating a practical problem which is usually expressed in words into arithmetical or algebraic terms.
2. Choosing a suitable mathematical technique to help us get the information we need (e.g. do we draw a graph, solve an equation, or what?)
3. Re-interpreting the solution which we obtain by mathematical methods in terms which are useful in practice.

We do not intend to go into this process in any great detail; our objective is simply to start you thinking about the ways in which even the elementary mathematics which we have been revising in

FIGURE 6.1
Using mathematics to solve problems

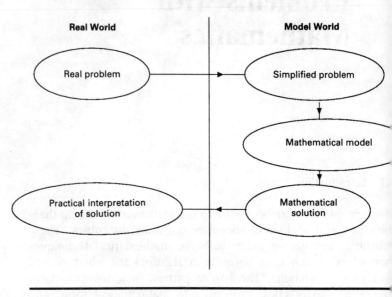

earlier chapters can be used to help solve problems arising out of real situations.

So in the remainder of this chapter we will try to offer some principles to guide you in tackling problems using quantitative methods. We say 'principles' because it is not really possible to specify a standard routine to be followed; this may be why novices find the whole process quite tricky at first. Instead, we will solve a number of problems in detail, drawing on the mathematics covered earlier in the book, and then reflecting on the steps we have carried out in each case.

We have, of course, already been applying our mathematics to practical problem-solving throughout the book: in Chapter 3 we investigated how linear demand relationships could be expressed mathematically and graphically, and how problems involving interest could be solved with the aid of functional expressions. In Chapter 5 we used calculus to help us maximise revenue and profits. So if you followed the arguments in those chapters, you would not find anything too surprising in this one.

There is one important difference, however, between our approach to problem-solving here and that in earlier chapters: here we were primarily concerned with the particular method under discussion – solution of quadratic equations, differentiation of linear functions, or whatever. In this chapter we are much more concerned with the *process* involved, so we would like you to try to concentrate on that rather than on the details of the methods we use.

For the same reason, we do not intend to demonstrate methods for solving complex business problems. If we did so, you might find yourself unable to see the 'wood' – that is, the general problem-solving direction – for the 'trees' of detailed mathematics. So we have chosen to illustrate our argument by using a number of fairly simple methods, which nevertheless are widely used in practice.

We begin by examining a problem related to questions about compound interest which have already been mentioned. From our approach to this problem, we extract a framework of steps which could be followed in solving any practical problem using mathematics: these are laid out in section 6.3. We then apply these steps to two further problems in the remainder of the chapter.

2 A Financial Problem

Suppose that a company knows it will need to replace a piece of capital equipment in three years' time, and wishes to make provision for this purchase by building up a special investment account. This section shows how mathematics is used to answer the following question:

> How much should the company invest per year to provide for the purchase of the replacement equipment, taking into account the interest which annual investments towards the equipment will earn?

Our first step is to *identify what information we can obtain about the problem*. We already know that the equipment will be replaced in three years' time'. We shall suppose that when we say 'three years' time' we mean precisely that – exactly three years from today. Next we need to know what the equipment will cost. Enquiries among equipment manufacturers suggests that a likely price will be £600 000.

You can see that here we have had to do something which is very common in applying quantitative methods to real problems: *make some not totally realistic assumptions* in order to simplify the problem. It is unlikely that the estimated price of £600 000 will turn out to be exactly right, or that we really will buy the equipment in precisely three years' time. However, we can think later about the impact of those assumptions on the validity of our solution.

Now we have to make another, more sweeping assumption: that we know what the interest rate available on invested capital is, say, 9% gross – and that it will not change during the next three years. This may well be fairly improbable, but in the absence of a crystal ball it is the best we can do. Our final assumption is that payments into the investment account will be made at year intervals, the first being made today, and the last one year before the equipment purchase. So there will be three payments altogether.

It is clear that, generally speaking, you need to think through your problem carefully in order to extract the information needed for a mathematical solution – and that some simplifying assumptions may be necessary initially.

Our next step is to translate the problem into a piece of mathematics, using the information we have gathered. We begin by *identifying the unknown quantity(ies) and giving it (them) names*. The single unknown quantity whose value we are seeking in this problem is the annual investment. If we want to be able to handle this quantity in mathematical expressions we need to give it a name. We might say, as we often do with unknown quantities, 'Let's call it y'. But since y is an amount of money, we really need to be more careful about how we define it. So we prefer to say:

let y = annual investment in £.

Having decided that y will be in £ – a sensible decision since the amounts of money involved are quite large – we need to make sure that any other amounts are in the same units. We could not, for example, decide to measure the cost of the equipment in hundreds of thousands of pounds and therefore call it just 6. This point is important enough to give us another general principle: *specify units clearly and make sure they are used consistently.*

Having defined our variable, we can now go ahead and translate the problem into mathematical form. Rather than trying to leap to the final equation in one go, we take it step by step. Start by asking yourself *'What basic relationship do I wish to express here?'*

In the investment problem the answer is:

Total of three annual investments plus interest earned = cost of equipment.

The cost of the equipment is easy – that is just £600 000. The total of investments plus interest is not quite so simple. Recall that in discussing compound interest on an investment in section 3.2, we found that the value of £1000 invested for t years at 10% per annum was:

$$1000(1.1)^t = 1000(1 + 10/100)^t.$$

This can easily be generalised to state that if £C is invested for t years at r% per annum, it will amount to:

$$C(1 + r/100)^t. \tag{6.1}$$

The first payment of £y is made today, and so will have been in the account for the full three years when we need to withdraw it to pay for the equipment. So the compound interest formula (6.1) tells us that by then it will amount to $y \times 1.09^3$. The second payment, however, will only have been invested for two years, and so will amount to $y \times 1.09^2$, while the third after only a year in the account will be just $y \times 1.09$.

Altogether, then, total amount invested plus interest at the end of three years = $y \times 1.09^3 + y \times 1.09^2 + y \times 1.09$, which can be simplified to $3.57313y$. (We need to retain plenty of decimal places at this stage, to ensure an accurate answer when the result is combined with a large figure such as 600 000.)

If we equate this to the cost of the equipment, we obtain the equation:

$$3.57313y = 600\,000.$$

We have now *expressed the relationship between known and unknown quantities in algebraic terms*. You should recognise this equation as a linear equation of the kind we solved in Chapter 4.

The next step is to *solve the equation*. Following the rules which you met in Chapter 4, you can check that:

$$y = 167\,920.$$

Finally we must carry out an important but often neglected step, namely *translate our answer into terms of the original problem* – which did not say 'Find y'! What we have found is that the annual investment is £167 920, so altogether we will pay £167 920 × 3 or £503 760 – the remainder of the £600 000 required will come from the interest.

In reality this should not quite be the final step, because we should ask ourselves one further question: *Does this answer look about right from a practical viewpoint?* For example, here we expect the yearly payments to be rather less than £200 000, which is what we would need to invest if there were no interest earned. So the figure we obtained seems credible. This kind of argument will not tell you if your answer is definitely exactly right, but it will give you an idea if you have made a major mistake such as keying in a wrong figure on your calculator.

If you want to be even more cautious, you can *carry out a check calculation*. Here we can follow through the life of the investment (generally known as a *sinking fund*) as follows, working to the nearest pound:

Year 1 investment	167 920
Year 1 interest @ 9%	15 113
Year 2 investment	167 920
Total at start year 2	350 953
Year 2 interest @ 9%	31 586
Year 3 investment	167 920
Total at start year 3	550 459
Year 3 interest @ 9%	49 541
Total at end year 3	600 000

We have gone into a lot of detail over this example, because we wanted to clarify the thinking behind each step. When you are used to handling this kind of process, you will not need to spell out every step as explicitly as we have done – some can be contracted or omitted altogether. (In fact there are tables and formulae available for dealing with sinking fund problems of this kind; some of you may encounter them in accounting courses.)

6.3 The General Problem-solving Process

If we extract from the last example the general principles which we laid down as we went along, we get the following list:

- **Identify what information we can obtain about the problem.**
- **Make some assumptions in order to simplify the problem.**
- **Identify the unknown quantity(ies) and give it (them) names.**
- **Specify units clearly and make sure they are used consistently.**
- **Ask 'What basic relationship do I wish to express here?'**
- **Express the relationship between known and unknown quantities in algebraic terms.**
- **Solve the equation.**
- **Translate the answer into terms of the original problem.**
- **Ask 'Does this answer look about right from a practical viewpoint?'**
- **Carry out a check calculation.**

We shall use this list as a guide when tackling the remaining problems in the chapter, and will see that, although not all the points will be relevant in every case, and some may need modification, the essential problem-solving structure is always much the same.

6.4 A Depreciation Problem

Let us suppose that a company has just purchased a new metalworking machine. The book value of this machine is to be reduced by equal amounts each year, until by the end of its useful life the value has decreased to its price as scrap. The problem we wish to solve is the following:

What is the general relationship between the age of the machine in years and its book value?

We begin, as before, by identifying the information we can obtain about the problem. The purchase price of the machine should be easily obtainable – let us agree that this is £13 000. The other important thing to find out is the expected useful life of the machine, and its scrap value at the end of its life. Here we will probably have to make some assumptions, based on experience with similar machines in the past. Suppose that we are told the machine can be expected to last eight years, after which it can be sold as scrap for £1500.

The major assumption in this problem has, of course, already been made – namely, that the value will decrease by equal amounts each year. This means that the rate of decrease of the value over time is constant, so that the graph of value against year would be a straight line, and the corresponding equation a linear equation such as we discussed in Chapters 3 and 4. The unknown quantity of interest is the value at any given point in time, which we will call £V. Notice that here we are using V for value, rather than the more usual x or y, to represent our unknown. It is often a good idea to represent unknowns by meaningful letters like this, especially when there are several variables to keep track of – we shall return to this point in the next problem.

The basic relationship we wish to express in this case is rather different from that in section 6.2. There we wanted to find the single value of an unknown quantity, so we needed to find an equation to solve. Here we have a more general requirement – to establish the relationship between V and time. If we use t to denote the time in years since purchase of the machine, then we require an equation linking V and t, which our assumptions tell us should be linear.

Now we know from Chapter 3 that a linear equation always has the form $y = ax + b$, where x and y are the variables; so in our case we want an equation of the form $V = at + b$. What we must do is to determine the values of the constants a and b which will give us the *right* line – that is, the one for which the value is £13 000 when $t = 0$, and £1500 when $t = 8$.

You know from Chapter 3 that the constant term b in this equation is just the value of V when $t = 0$. So we get $b = 13\,000$

immediately. That means that we can write the equation as $V = at + 13\,000$. The other point on the line which we know is $V = 1500$ at $t = 8$, which on using the equation gives:

$$1500 = 8a + 13\,000.$$

You should be able to solve this for a, obtaining:

$$a = -1437.5.$$

Thus the final equation is:

$$V = -1437.5t + 13\,000$$

which looks more familiar the other way round:

$$V = 13\,000 - 1437.5t.$$

The step 'solve the equation' in our outline scheme does not apply here. The equation cannot be solved, since it involves two unknowns, and in any case our objective was to obtain the relationship between V and t, not specific values. So we proceed to the next step, translating the result into practical terms.

The 13 000 gives us the value of V when $t = 0$, which of course, we already knew. The other constant is the slope, which is -1437.5 (do not forget that the sign forms part of the value). This can be interpreted as indicating that for every year we keep the machine, its value decreases by £1437.50. Another way of arriving at this figure would be to observe that the value decreases by £13 000 − £1500, or £11 500, over 8 years − that is, £1437.50 per year.

Finally, we ask whether this result seems sensible. Certainly the negative slope is not unexpected − value decreases as time increases, so we should have been prepared for this. The equation would, of course, give a negative value of V when, say, $t = 10$, but then we only intend it to be used over the lifetime of the machine, which is eight years. Within that range, the equation looks reasonable.

Some of you may have recognised this as a problem concerned with depreciation, other examples of which you met in Exercises

3.1 and 3.2. We have used the simplest method, called *straight-line depreciation*. However, you can use the same sort of argument to obtain expressions for more complicated depreciation laws, though if you want to fit a curve, such as exponentially decreasing value, to the data, you may need to know more than two points on the curve.

6.5 A Stock-control Problem

Our final example relates to the cost of obtaining and storing goods. Most organisations have to do this, whether they are supermarkets obtaining supplies of tinned goods from a warehouse, or steel-mills obtaining raw materials to go into the steel production process. Management then needs to decide how large a quantity should be ordered at one time – is it better to order large supplies at infrequent intervals, or small amounts more often? We would like to construct a mathematical model to help in answering this question.

Compared with our earlier problems, this one seems rather ill-defined. However, a little thought shows that what we really want to know is how to determine the order size which makes costs as low as possible – in other words, to *minimise total costs*. After your work in Chapter 5, the word 'minimise' should immediately cause you to think of calculus – and indeed, we are going to use differentiation to help find the cost-minimising size of order (though if you have not worked through Chapter 5, do not worry – we shall also be pointing out alternative ways of arriving at the solution).

Before we can start trying to minimise the costs, we need to derive an equation showing how they relate to the size of the orders. Our first step, as in both the earlier problems, is to identify known information about the problem.

Historic records will probably enable us to estimate the current demand for the item – let us say it is 12 000 items per year. We have to start making assumptions immediately by assuming that this figure will not alter too much in the near future.

Then we move to the costs. There will be a cost associated with raising an order (administrative cost, delivery charge and so on). With the aid of an accountant we should be able to get an idea of this cost – suppose it is in the region of £40. Now we need another assumption: order cost does not vary with the size of the order. (Since we are aiming to help you learn mathematics rather than operations management, we shall not spend time considering how likely the assumptions are to be true, but you might like to give this question some thought.)

The other element of cost will be connected with the storage of the items; if we order in bulk, we will have a lot more items to store than if we order small amounts at a time. Again, we need to estimate this storage or stockholding cost; we might discover that to store one item for one month costs 2p.

Following the usual procedure, we next decide what the unknown quantities are, and what we are going to call them. There are two things in the problem which may vary – we may choose to change the order size, and the total cost of placing orders will vary as a result. Making sure we include the units in our definition, we say:

let C = total cost in £ per annum; and
let q = number of units per individual order.

As before, we use meaningful letters rather than the ubiquitous x and y.

At this point we consider the basic relationship we are trying to express; here it is quite simply stated in words:

total cost = ordering cost plus stockholding cost.

Translating this into an algebraic equation is not quite so simple, and needs to be broken down into manageable steps. The LHS is no problem: total cost is what we have called C. On the RHS, the first term is ordering cost, which will depend on how many orders we place per year. We know that demand is 12 000 items per year, and we also know that the quantity being ordered at a time is q. But how can we link these to give us the number of orders in a year?

You may be able to see immediately that the number of orders will be $12\,000/q$. If making the leap directly to this formula seems difficult, think about what would happen if q were 3000 items; then you would need 4 orders to last the year. Similarly, if q were 4000, the number of orders would be 3. So to find the number of orders per year, we are dividing the size of each order into the total demand of 12 000. Thus for a general size of order q, the number of orders would be $12\,000/q$.

What we have done here is so useful that it almost deserves the status of another general principle:

- **Try out calculations with simple numbers before attempting to express them in general algebraic terms.**

Having found the number of orders per year, we just multiply it by the cost of each order to find the annual ordering cost. So altogether,

annual ordering cost in £ = $12\,000/q \times 40 = 480\,000/q$.

The other contributor to total cost is the stockholding cost. This is 2p per month for each item held – but the number of items being held will vary as orders are delivered and used up. To get a definite expression for stockholding cost, we need a further assumption: stocks are used up at a steady rate from a maximum of q (just after an order is delivered) to a minimum of zero. At this point there is instant replenishment of the stock, and the whole process starts over again. Figure 6.2 is a graph showing how stocks vary with time if this assumption is made.

If you accept this assumption (which, we agree, takes some swallowing, especially the 'instant replenishment' part) then it follows that in the long term the average amount in stock will be $q/2$. (If you cannot see this immediately, note that this is the stock level half way through the period between replenishments; higher stock levels early on in the period are balanced exactly by lower stocks later in the period.) Since each of the $q/2$ items costs 2p to store for a month, the stockholding cost will be $q/2 \times 2$ pence per month.

However, we need to be careful again here about the units. We already decided to express total cost C in £ per year, so we cannot have stockholding cost in pence per month. Making the change to £ per year gives $q/2 \times 2 \times 12/100$ or $0.12q$ £ per year.

Now at last we can write down the complete equation relating C and q; it is:

FIGURE 6.2
Graph of idealised stock cycle

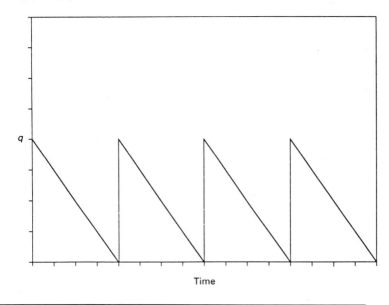

C = order cost + stockholding cost

$= 480\,000/q + 0.12q.$ (6.2)

You may have lost sight of the original objective in the lengthy process of obtaining this equation; what we actually set out to do was to find the size of order which makes cost a minimum. Rephrasing this in terms of our mathematical expression of the problem: we want to determine the value of q (if any) which minimises C.

We have said 'if any', but a look at equation (6.2) should show that there certainly will be a minimum value of C. As q gets bigger, the first term – the order cost – gets smaller; a sensible result, since with bigger orders we need to order less frequently. On the other hand, the second term – the stockholding cost – grows linearly with q. So there must be a point representing a 'happy medium', where the two costs balance out. This will be our minimum cost policy.

We have a choice of methods for determining the value of q which produces this minimum cost. We can use calculus and say that C will have a minimum (or maximum) when $dC/dq = 0$. Applying the differentiation rules of Chapter 5 we get:

$$dC/dq = -480\,000/q^2 + 0.12$$

and this is zero when

$$-480\,000/q^2 + 0.12 = 0,$$

which solves to give $q = +/-2000$. Clearly a negative value of q makes no practical sense, so the value we are looking for must be $+2000$. If you want some practice in differentiation, you can find the second derivative of C with respect to q, and show that it is positive when $q = 2000$, so that C is indeed a minimum at this point.

The alternative method, if you do not want to use calculus, is to plot a graph of C against q and use this to read off the value of q making C a minimum. You should recognise that the graph will be a curve, since its equation contains a $1/q$ term. So you will need to plot quite a few points to get an accurate picture; the realisation that annual demand is measured in thousands should suggest that values of q of the order of several hundred, rather than $q = 1, 2, 3, \ldots$, are sensible ones to choose.

Whatever means we use to find the best value of q, we need to translate the solution back into a practically useful form. We have found that ordering 2000 items at a time minimises the total cost of ordering and storing them. Given that 12 000 items a year are needed, this means that six orders will need to be placed each year – that is, one every two months.

The problem we have examined in this section is an example of an *inventory problem*. In the Exercises at the end of the chapter, you will find a more general problem of the same type, the solution of which gives a formula for determining the cost-minimising order quantity for any demand and cost structure.

6.6 Conclusion

We observed at the start of this chapter that there are really no standard 'methods' for applying mathematical techniques to the solution of real problems. Nevertheless, we hope that the framework of general principles developed and illustrated in this chapter will stand you in good stead in applying the mathematical and statistical methods you learn in the future. Perhaps the most important principle of all is to take things steadily, step by step, and never to let go of your common sense!

Exercises 6.1

1. You know that five years from now you will need to buy a replacement car costing £8000. Interest is currently available at 11.5% per annum (compounded annually). How much do you need to invest now as a single lump sum, in order to accumulate the necessary amount by the end of the five years? What assumptions do you need to make to reach your solution, and how realistic are they?
   (The answer you obtain – assuming it is correct! – is what is known as the *present value* of a sum of £8000 payable in six years at a *discount rate* of 11.5% per annum. If you take courses in accounting or finance, you will learn about the use of net present values as one method of investment appraisal).

2. Suppose that in the inventory problem of section 6.5, the annual demand was for D items, the cost of placing one order was £R and the cost of holding a single item in stock for a year was £H. Follow through the argument of section 6.5 using these general values for the quantities, and thus obtain a formula giving the cost-minimising quantity q in terms of R, H and D.

3. (a) Use the formula you have derived in question 2 to determine the cost-minimising quantity to order if the demand for an item is 200 per month, the cost of placing an order is £4 (irrespective of size), and the cost of holding an item in stock for a year is 12p. How often will you need to place an order?

(b) Suppose you subsequently find out that your estimate of the order cost is wrong, and in fact it should have been £3.50. What percentage error in the cost have you made? What should the correct order size be, and how much extra per year will you be paying in inventory charges as a result of your error? What general implications does your conclusion have for the usefulness of this method?

4. When a price of 40 pence per unit is charged for a commodity, annual demand for the commodity is 20 000 units. If the price is increased to 50 pence per unit, demand falls to 18 000 units per annum.
   (a) Assuming that demand varies linearly with price, determine the equation linking them.
   (b) Hence find at what price the demand would fall to zero. How likely do you think this is in practice? Where might the linear model break down?
   (c) Also determine what the demand would be if the commodity were free of charge.

5. A direct-sales firm uses the following rule-of-thumb for determining postage and packing charges: 5% of order value for orders up to £50, 3% for orders above £50 but below £75, and free of charge thereafter. Devise a chart from which staff can easily read off the charge for any given order value.

7 Simple Statistics

7.1 What is Statistics About?

So far in this book we have been looking at various concepts and techniques which could be classed under the general heading of mathematics. For this last chapter, however, we are moving into a slightly different area – that of statistics. While closely related to mathematics, and using many mathematical ideas and methods, statistics addresses a distinct set of problems, and so it is worth spending a little time discussing what those problems are.

Throughout business and the social sciences, progress is very dependent on having access to information or *data*. To take a few illustrations:

- Before launching a new product, a company will carry out market research to determine the sales potential of the product, where it should be sold, what type of advertising would be most effective, and so on. The initial result of this research will be a large amount of data, probably in the form of answers to questionnaires, which will need processing before it can be used to help with decision-making.
- In order to keep track of the operation of a large automated plant such as a steel mill or chemical works, many measurements will be taken at regular intervals on various aspects of the plant (pressure in reactor vessels, temperature of waste gases, amounts of chemical input into the process, etc). The data resulting from this monitoring process will almost certainly be gathered by computer, but it must still be processed in a suitable way if it is to provide vital information about safety, economy and so on.
- A psychologist conducting research into aspects of children's development may conduct various experiments, perhaps observing the children playing with specially-designed equipment,

noting how many times they carry out a particular movement, and so on. Such experiments generate a mass of data, which cannot lead to useful conclusions unless it is analysed carefully.

In all three of these examples there is a need to process a large amount of data in some way so that it becomes more informative. This is the essential problem to which statistics addresses itself. You have only to reflect on your own experience to realise how necessary this processing is: what would be your reaction if, applying for a new job and enquiring about the likely salary you would receive, you were handed a set of payslips for the last month and told to draw your own conclusions? You would probably not be best pleased! – what you want is a summary of the information, indicating starting levels for different ages and qualifications, annual increments, and so on.

Statistics, then, is concerned with methods for gathering, summarising and drawing conclusions from data. Generally, this data is either already quantitative (age, annual salary, temperature, pressure, weight, . . .) or can be quantified to some extent ('rate the smell of this product from 1 (very unpleasant) to 5 (very pleasant)'). Even when the data is apparently purely qualitative (gender, region, . . .) we can often get some numbers out of it by doing counts (How many males and females in the group? How many sales outlets in each region?).

We cannot cover a great range of statistical techniques in a book of this nature; in any case, if you are about to study quantitative methods as part of a business or social science course, a large part of those studies will probably be devoted to statistics. However, since many school maths syllabuses now include some basic statistics, and since it will be an advantage to you to be familiar with simple methods for summarising data before you start your course, this chapter aims to introduce you to a few of those methods.

7.2 How to Organise Statistical Data

7.2.1 *Types of data*

We have already seen that data comes in different forms, some more apparently quantitative than others. We need to think about this in a bit more detail before we examine methods for organising the data.

Counts, like the numbers of males and females in a group of workers, provide an example of what is called *discrete* data. We might define this as data which can only take certain definite values; a count, for instance, has to be a whole number.

Continuous data, on the other hand, can take absolutely any value (though the values may be confined within a limited range). The pressure in a chemical reactor would be an example of a continuous quantity – it can vary over a whole range of possibilities, though presumably if it goes outside that range there will be problems – the reaction may come to a halt, or the vessel explode!

We also need to distinguish different sources of data. We talk about a *population* if we are gathering data from every single item, person or whatever that is of interest to us. So if we weigh every packet of detergent which comes off an automatic packing line, then we have an entire population of measurements. The process of gathering data from every member of a population is sometimes called a *census* (like the familiar ten-yearly census of the British population).

If, on the other hand, as often happens, we cannot gather data from every item of interest (perhaps because it would be too expensive, or take too much time) then we have to content ourselves with a *sample* of data. Most statistical measurements are effectively based on samples; clearly, the way a sample is selected will have a big effect on the reliability of the measurements, and so later in your studies you may discuss sample design in detail. Meanwhile, as we shall see in section 7.4, there are certain technical differences which depend on whether your data is derived from a sample or a population.

We often need a way of referring to the quantities which compose our set of data; a useful term is *variable*, which is used to denote any measurement which varies from one member of our population or sample to the next. So *height* is a quantitative variable which could be measured for a group of people; *region of residence* is a qualitative variable which could be obtained for the same group. Statistical variables of this kind are not used in quite the same sense as the algebraic variables you met in Chapter 2; however, one thing they have in common is that they are often denoted by letters such as x or y.

7.2.2 Tabulating statistical data

Confronted with a set of individual responses to the question, 'Please indicate whether you are male or female' most people would have no difficulty in deciding to summarise the data by means of a table in the form:

Gender	No. of respondents
Male	36
Female	44

The same kind of idea can be easily applied to most qualitative variables. The usefulness of the table is often enhanced by including extra information such as a total, and perhaps percentages indicating what proportion of the group fell into each category:

Gender	No. of respondents	Percentage
Male	36	45
Female	44	55
Total	80	100

Notice how the total and the percentages have been quite clearly distinguished from the actual figures in the table, so as not to create confusion.

A final refinement is to add a title and indicate the source of the data (for anyone who wants to go back and look at it in more detail):

Gender of shoppers at XYZ supermarket

Gender	No. of respondents	Percentage
Male	36	45
Female	44	55
Total	80	100

(*Source*: Customer survey, 19/8/92)

The figures in the 'number of respondents' column are generally called *frequencies*, since they indicate how frequently values in each category occurred. The letter f is often used to denote frequency.

This tabulation idea can be extended to data which has been categorised according to more than one variable. For example, if in the survey mentioned above the customers had also been asked, 'How long on average do you spend on each visit to XYZ supermarket?', then the responses to this question could be cross-tabulated with the gender responses, giving:

Cross-tabulation of gender against length of visit
Length of visit (minutes)

Gender	< 10	10 but < 30	30 and more	Total
Male	15	11	10	36
Female	8	22	14	44
Total	23	33	24	80

We could add percentages to this table, but we now have a choice of showing the percentage of the overall total, the percentage of the row (e.g. what % of males spent less than 10 minutes) or the percentage of the column (e.g. what % of people spending 30 minutes or more were female). Which version we choose depends on which aspect of the data we wish to emphasise. You may want to experiment with different versions of a table until you find the one which best suits your needs – the availability of computer software to do the tabulating makes this very much quicker than it used to be.

There are numerous general principles which can be given for constructing good tables, to be found in many elementary statistics books – things such as making sure that your categories do not overlap, arranging categories in a logical order, and so on. Most of these, however, are really just common sense; whenever you have drawn up a table, try to stand back and ask yourself whether someone looking at it for the first time could understand the content quickly without any further explanation from you.

You can use your ingenuity to construct tables showing more than two aspects of the data at once, but these can get confusing, and generally it is preferable to use several separate tables rather than one big one crammed with information.

7.2.3 Frequency tables

In the last example, you may have noticed that the categories for 'length of visit' were not qualitative, but were in fact groupings of a quantitative variable. This is how we usually deal with quantitative variables. Sometimes, if the variable is discrete, we can give an exhaustive listing of the possible values:

No. of visits to XYZ supermarket in the last month	No. of respondents
1	15
2	19
3	22
4	20
5 or more	4
Total	80

Even here, although we can list all the discrete possibilities, it is more efficient to include an open class (5 or more) at the end rather than trying to list all the possibilities up to the one person who has made 14 visits. A table of this form is known as a *frequency table*, or sometimes a *frequency distribution* since it tells us how the frequencies are distributed between the classes.

When the range of a discrete variable is larger, we may prefer to construct a *grouped frequency table*:

No. of items purchased on this visit	No. of respondents
1–5	6
6–10	15
11–15	37
16–20	12
21–30	7
31 or more	3

Notice how all the classes do not have to be the same width; note too that although there appear to be gaps between the classes, in practice they do not matter because, this being a discrete variable, there is no possibility of encountering a value which would 'fall 'down' one of the gaps – it is impossible for someone to have purchased 20.6 items. And this format with gaps is preferable to using 1–5, 5–10, etc., where there is ambiguity about exactly which class 5 items would belong to.

With a continuous variable, however, we need to be a bit more careful. Let us look again at the rather coarse grouping of the length of visit figures given above:

Length of visit (mins)	No. of respondents
< 10	23
10 but < 30	33
30 and more	24

Because time is a continuous variable, we cannot leave any gaps between the classes – absolutely any positive value of time could occur in practice. So we have adopted the clumsy-sounding but safe 'x but $< y$' format to overcome the problem without introducing any ambiguity. This is not the only way of dealing with such a situation; if you are familiar with other systems for specifying class boundaries, by all means go on using them, as long as you are clear about the principles. Generally speaking, about 6–10 classes are regarded as a good number to aim at, giving a fair degree of simplification while avoiding excessive computation.

It is sometimes more convenient to use an alternative format, the *cumulative* frequency table. This most often takes the form:

Length of visit (mins)	No. of respondents (cumulative frequency)
< 10	23
< 30	56
< 60	80

though a 'greater than' version is also sometimes used. Notice how in order to complete the table we have had to make an assumption that no one spent more than an hour in the supermarket. As we go through the table we accumulate more and more of the frequencies, hence the name cumulative. If the addition is correct, the final figure in the 'cumulative frequency' column should be equal to the total number in our sample.

Whichever format of table you choose, it is going to be necessary to count the number of items falling into each class. If you have studied this topic before, you may have spent time learning how to construct a 'tally' for this purpose; however, except for very small samples it is likely that these days you would be using some kind of computer software to actually construct the table. However, decisions on layout, number and range of classes, and so on still need to be made by the user, so you cannot abandon the whole exercise to the computer!

Exercises 7.1

1. Convert the following frequency table to a 'less than' cumulative format:

Mark (%)	Number of students
under 40	3
40 but < 50	8
50 but < 60	11
60 but < 70	25
70 and over	9

2. Construct a table to show the information contained in the following report: 'The distribution of degree classes obtained by students on the BSc Herpetology degree this year differs quite a lot from that for last year's group. This year, out of 27 students, only 1 obtained a First, but there were 18 2.1s and no Thirds or passes at all. Last year, in contrast, although there were 3 Firsts among 25 students, there were also 1 Third and 1 Pass, the remainder being split 60:40 between 2.1s and 2.2s.'

    (*Note*: In answering this question, you need to know that British degree results are divided into six categories, in decreasing order of standard: First; Second Class Division I (2.1 for short); Second Class Division II (2.2); Third; Pass; and Fail.)

3. Sketch two alternative frameworks for a table in which you could show simultaneously the breakdown of a group of 64 retail stores into

   (a) North/Midlands/South sales regions;
   (b) High Street/shopping precinct sites;
   (c) Profitable/loss-making last year.

7.3 Statistical Diagrams

7.3.1 *Diagrams for qualitative data*

Most people find that it is much easier to understand information if some kind of pictorial or diagrammatic representation accompanies it. We have already used this idea in drawing graphs to enhance our understanding of functional relationships. In this section we will look at some of the diagrams which can be used to represent statistical data.

(a) Pie-charts

The concept of the pie-chart is very simple, which is probably why it has become one of the most popular ways of presenting statistical information. A basic circle (the pie) represents the total group under consideration, and this is subdivided into sectors (or slices) in proportion to the breakdown of the group into subgroups. So the information concerning gender of shoppers in the XYZ survey, mentioned in section 7.2, would give the pie-chart shown in Figure 7.1.

If you need to calculate the angles of the 'slices' by hand, the easiest way is to work from the percentage figures calculated above. We saw that 45% of the shoppers were male. However, the total angle in the centre of the circle is 360°, so the angle of the slice representing males should be 45% of 360°, or 162°.

The slices of the pie should if possible be arranged in decreasing order of size. It is not a good idea to use a pie-chart when there are a great many categories in the data, particularly if some of them are quite small; Figure 7.2 shows how cluttered such a chart can become.

FIGURE 7.1
Pie-chart of gender data

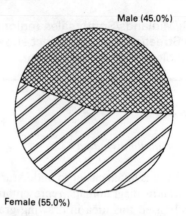

FIGURE 7.2
An overcrowded pie-chart

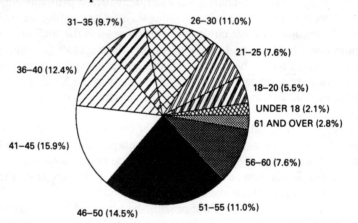

There are also problems in using two or more pie-charts to show the breakdown of different totals, since, strictly speaking, the sizes of the pies should be proportional to the totals involved. Thus if we want to show that last year £2.1m was distributed as dividends out of a total of £8.9m profit, whereas this year the comparable figures are £1.8m and £7.3m, the ratio of the areas of the two pies should be 8.9:7.3. If you recall that we find the area of a circle from the formula πr^2, you can see that finding the values of r to give the correct ratio will not be much fun! One way to avoid this is to show the percentage breakdown for each year, in which case both pies, representing 100%, can justifiably be the same size – but we then sacrifice the information about the *actual* profits.

To summarise: a pie-chart is best used to show the breakdown of a single total into not too many categories. Software packages now available enable you to construct very attractive pie-charts incorporating different colours or shadings for sectors, 'exploding' of one or more sectors, and explanatory labels.

(b) Bar-charts

A bar-chart offers more flexibility than a pie-chart as a way of presenting qualitative information, while being possibly a little less eye-catching. The bar-chart in Figure 7.3, which shows the data on gender/length of visit from the XYZ survey, illustrates the main advantages of the format.

We have chosen to show the gender as the major bars, with a subdivision of each bar by time spent; the chart could, of course, be drawn with time as the major division – the choice, as with tabulation, depends very much on the user's needs. Notice how spaces separate the bars, and how the width of the bars is purely a matter of convenience; it has no numerical significance. Only the height of the bar – representing the frequency within the group – is important. We hope it hardly needs to be said that, however tempting it may be in terms of saving paper, it is *never* correct to start the vertical axis of the chart from any value other than zero. (To see why this is, try putting a sheet of paper across Figure 7.3 at various levels, and notice how the proportionate heights of the bars are distorted.)

We could have chosen to place the 'time spent' components of the chart side-by-side rather than on top of each other, as in Figure 7.4.

FIGURE 7.3
A bar-chart

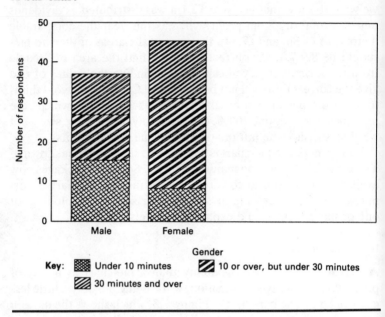

Or we could instead adopt a percentage-based format (though, as usual, with the corresponding loss of information about the actual totals) – see Figure 7.5.

From these examples you can see that bar-charts offer a great deal of scope to the ingenious user in presenting information concisely yet accurately.

7.3.2 *Diagrams for quantitative data*

(a) *Histograms*

Suppose that the XYZ supermarket survey also included a question about the distance travelled by shoppers to reach the store. The resulting data has been summarised in the following frequency table; you can see that no one in the sample travelled more than five miles:

FIGURE 7.4
Alternative form of bar-chart

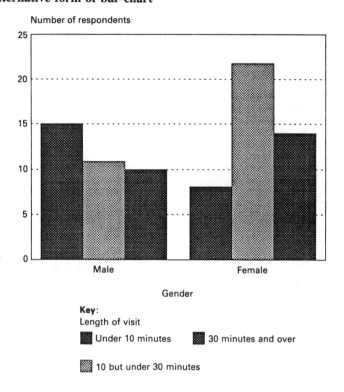

Distance travelled (miles)	No. of respondents
Up to 1	12
1 but under 2	16
2 but under 3	30
3 but under 4	13
4 but under 5	9

The usual way of showing this kind of data diagrammatically is by means of a *histogram*, as shown in Figure 7.6. As you can see, in many ways the histogram resembles a bar-chart, with the height of

FIGURE 7.5
A percentage bar-chart

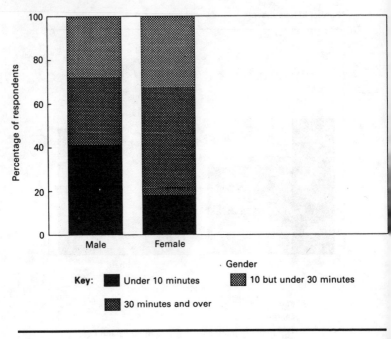

each block in proportion to frequency. However, there are no gaps between the blocks, since we now have a continuous scale of distance on the horizontal axis.

There is a more important distinction between a histogram and a bar-chart which is not apparent from this example, since all the classes here happened to be of the same width. Perhaps the best way to see this distinction is to examine Figure 7.7, which shows the 'length of visit' data plotted as a histogram. In drawing this histogram, we have had to assume that the longest time spent by anyone in the supermarket is 60 minutes – otherwise it would have been impossible to show the top class.

If you now compare Figure 7.7 with a bar-chart such as Figure 7.3, you will notice that the bars in Figure 7.7, unlike those of the bar-chart, are unequal in width, so that the unequal time-ranges are correctly represented. As a consequence, the vertical axis does *not* represent the frequency.

FIGURE 7.6
Histogram of distance travelled

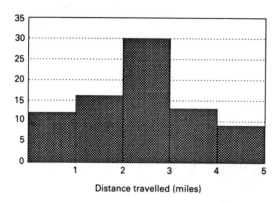

FIGURE 7.7
Histogram with unequal classes

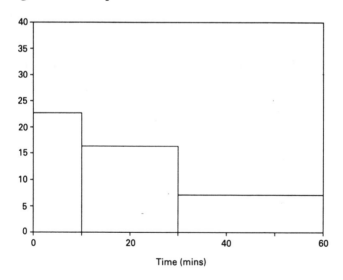

FIGURE 7.8
Histogram wrongly plotted

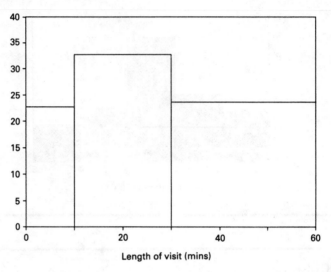

To see why this must be, observe that what impresses the reader when looking at the diagram is the area occupied by each of the bars. Thus in view of the unequal width of the classes, heights must be adjusted so that the area gives a correct impression. If the first class, with a width of 10 minutes, and a frequency of 23, is represented by a bar with a height of 23 units, then the second, which is twice as wide, need only have a height of 33/2 or 16.5 to show the frequency of 33 correctly. Finally the last class, width three times as great as the first, and with a frequency of 24, has to be given a height of $24/3 = 8$.

If this seems unnecessarily complicated, compare the impression given by Figure 7.7 with that produced by Figure 7.8, where the bars have been simple-mindedly plotted with height proportional to frequency. The overall impression produced by the diagram is that there is a very large number of people spending a long time in the supermarket – a fact which we know is not true.

To avoid the difficulties created by unequal classes, you can, of course, when given the choice always opt for equal classes as far as possible. If you do so, then heights of bars *will* be proportional to

FIGURE 7.9
Ogive of length of visit data

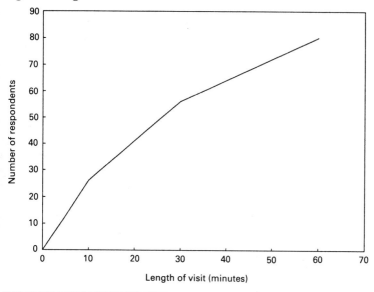

frequency, so there is no problem. You should, however, be aware that many software packages are unable to cope with histograms having unequal classes – all bars will be drawn the same width, and some widely-used packages invariably leave gaps between bars, thus effectively giving a bar-chart even when a histogram is what is required.

(b) Ogives (cumulative frequency graphs)

The histogram is obtained from the ordinary frequency table. If instead you want a diagram to show cumulative frequency information, then you need an *ogive*, which is a great deal easier to plot than a histogram. Look at the ogive in Figure 7.9 for the 'length of visit' data – what we have is simply an ordinary graph, showing the number of customers who spent less than a given length of time at the supermarket. Joining the points of the ogive with straight lines makes the assumption that values are evenly spread within the classes (for example, that of the 24 customers spending between 30 and 60 minutes, 12 spent under 45 minutes and the remainder spent between 45 and 60 minutes). This may or may not be the case, but in the absence of more detailed information it is probably the most reasonable assumption.

Having plotted the ogive, we can use it to read off intermediate values; for example, about 31 customers spent 15 minutes or less on their visit to XYZ. We will see in the next section that this can be very useful when we come to finding summary measures for the data.

7.3.3 Other statistical diagrams

There are many other types of statistical diagram which you may come across – *pictograms* (using scales such as 'one drawing of a little person = 100 jobs lost'), *statistical maps*, and so on. Keep your eyes open when reading books and articles for imaginative uses of graphic presentation of data. As far as drawing your own diagrams is concerned, the same warning applies as with tabulation: ask yourself whether a reader, looking at the diagram for the first time, would obtain the impression you really want him or her to get. If not, think again!

Exercises 7.2

1. Plot a histogram and an ogive for the data of Exercises 7.1, Question 1. Use your ogive to determine how many students pass the examination, if the pass mark is 45%.
2. Construct a suitable diagram to represent the data contained in the report shown in Exercises 7.1, Question 2.

7.4 Measures of Location and Dispersion

7.4.1 The need for summary measures

The tables and diagrams we have constructed in sections 7.2 and 7.3 have gone a long way towards simplifying the presentation of data, but for many purposes we want to go even further, and sum up the important facts about a set of quantitative data via just two or three numbers. So what are the important features of the data we might wish to measure? We can get some ideas by looking back at the histograms we have drawn – the one in Figure 7.7, say for the 'length of visit' data.

First, it would be useful to have a figure to tell us what a *typical* value from the distribution is. This will give us an idea of what sort of figures we are talking about overall. From Figure 7.7 it looks as if this 'typical' figure will be somewhere in the 10–30 minute range. Such a figure is often referred to as a measure of the *location* of the distribution.

Second, it would be helpful to know how the individual values in the distribution are scattered – are they very spread out, or are they all quite close together? The values shown in Figure 7.7 range between 0 and 60 minutes, but it might be that had we done our research at a busier time, the range would increase to 0 to 90 minutes or even more. We call measures which address this feature of the data measures of *dispersion*.

A third aspect which is sometimes measured is the symmetry or otherwise of the data – Figure 7.7, for instance, is quite unsymmetrical or *skewed*. However, this is a less important issue, and we shall not be pursuing it here.

7.4.2 *Measuring location*

There is no one universally accepted way of measuring location (or, for that matter, dispersion), since different types of data call for different measures, as we shall see. We are going to concentrate on the three measures which you are most likely to come across in quantitative methods courses – and which you may already have met in your school-level studies.

(a) *The (arithmetic) mean*

We have put the word 'arithmetic' in brackets in the title of this section, because, strictly speaking, it should be there, since other kinds of means can also be calculated. In practice, however, the arithmetic mean is by far the most common, so on the whole if someone mentions a mean you can take it that this is the one they are using.

The mean is nothing other than the old familiar 'average' which you have known how to compute since primary school days. It can be found very easily if we actually have a small set of individual values; for instance, if five men have weights in pounds of 120, 131, 147, 152 and 160, then their mean weight is (120 + 131 + 147 + 152 + 160)/5 = 142 lb.

The mean has many advantages as a measure of location. It has a simple interpretation, if a rather gruesome one in this case: the mean weight would be the weight of each member of the group if their weights were combined and then shared out equally among the group (in other words, melted down and recast into five equal-sized people!). A consequent advantage of this fact is that, given the mean of a sample and its size, we can always recover the total for the sample: if we are told that the mean weight is 142 lb, we can deduce that total weight = 142 × 5 = 710.

A more important fact relates to the point we mentioned earlier, concerning the difference between data based on a sample and that based on a complete population. The mean of a sample gives a good and predictable guide to the mean of the population from which the sample came, assuming that the sample was selected in a 'sensible' way. If you subsequently follow a quantitative methods or statistics course, you will certainly spend some time discussing just *what* we can say about the population mean armed only with sample data.

However, there is one snag about the mean. It can be rendered less than useful if our set of data contains 'rogue' or extreme values. For example, if the sample of five men whom we weighed happened to include a sumo wrestler with a weight of 280 lb, in place of the 120 lb man, then the mean (check this!) would become 174 lb. This figure is not really typical, either of the four lighter men or of the one much heavier person. The same consideration, possibly in a less extreme form, applies whenever we have a very skew (unsymmetrical) set of data. So in such a situation we might want to try a different kind of measure (see below).

Now let us turn to how the mean is actually calculated. It is simple enough, as we have seen, if we still have the raw data – but suppose we have thrown that away, and retained only a summary frequency table. What if, for instance, we needed to find the mean number of items purchased by shoppers at the XYZ supermarket?

You will recall that the data on this aspect of the XYZ survey was summarised in section 7.2 as follows:

No. of items purchased on this visit	No. of respondents
1–5	6
6–10	15
11–15	37
16–20	12
21–30	7
31 or more	3

So in principle, what we want to do is to add up the numbers purchased by the 6 people in the 1–5 range, the 15 in the 6–10 range, etc., and then divide by the total of 80 shoppers in the sample. There is just one problem: we do not know exactly how many items those 6, 15, . . . people purchased – all we know is the ranges into which they fell. So we have to make an assumption, and pick a single value to represent each class. The most defensible assumption is to use the midpoint of the class as representative (found by adding together the class limits and dividing by two): thus we assume each of the first 6 people bought 3 items, each of the next 15 bought 8, and so on. The result we get will not, of course, be guaranteed to be the same as that which would result from using the original data, but the discrepancy should not be too great.

We also need to close the last class at a sensible value – we will use 50, though other values such as 40 would also be reasonable. We then have a set of variable values (denoted, as you will remember, by x), representing the numbers of items, and a corresponding set of frequencies (f). So the table now looks like this:

No. of items purchased on this visit	No. of respondents, f	Class midpoint, x
1–5	6	3
6–10	15	8
11–15	37	13
16–20	12	18
21–30	7	25.5
31 or more (50)	3	40.5

Some of the midpoints have turned out not to be integers, but that does not matter – they are only theoretical figures.

Now we want to add up the six values of 3, the 15 eights, and so on. But of course there is no need to do things so laboriously – saying 6 × 3 gives the same result as saying 3 + 3 + 3 + 3 + 3 + 3, and is a lot quicker! What we do, then, is to multiply each x by its corresponding f, and then add the whole lot together.

This is where a symbol we introduced back in Chapter 1 will come in useful. Remember the summation sign Σ, meaning 'add up'? Our process for finding the mean involves adding up all the fx products, so we can write this as Σfx. Finally, we want to divide by the total number of people in the group, which is equal to the sum of all the fs or Σf. So we obtain a general formula for getting the mean from a frequency table:

mean = $\sum fx / \sum f$.

Check that carrying out this process for the 'number of items' data gives you 14.1875 as the average number of items purchased. If you are using a calculator, use the memory to add up the fx values as you work them out; then what you get out of the memory at the end will be Σfx, and all you need do is divide it by 80. The table including the fx values is shown below so that you can check your calculations.

No. of items purchased on this visit	No. of respondents, f	Class midpoint, x	fx
1–5	6	3	18
6–10	15	8	120
11–15	37	13	481
16–20	12	18	216
21–30	7	25.5	178.5
31 or more (50)	3	40.5	121.5

Various letters are used to denote the mean, the most common being $\bar{x}$ (pronounced x-bar) for the mean of a sample, and μ (a Greek letter m for mean, called 'mu', pronounced mew) for the mean of a population. This convention of ordinary letters for sample values and Greek ones for population values is almost universal.

Even cheap calculators now often have the ability to calculate means automatically, but as there is no standard key-sequence for doing this, we cannot attempt to explain the process; you will have to read the explanatory leaflet for your own particular machine (we hope you have not lost it or thrown it away!). But whether you are obtaining your means via a calculator, a computer package, or 'from scratch', you still need to have an appreciation of the actual process as described here.

(b) The median

We have seen that for very skew distributions, the mean can be somewhat distorted. An alternative in such cases is provided by the *median*, which is the central value in the distribution when it is arranged in ascending or descending order. Take the case of our five men's weights, which were, in ascending order, 131, 147, 152, 160 and 280 lb. The median here would be 152 lb, and so is unaffected by the extreme 280 lb value.

In this case we had 5 values, of which the median was the third. In general, if we have n values the median will be the $(n + 1)/2$th. If n is an even number, this formula gives a fraction – with $n = 10$, for example, we would find the median at the 11/2th or 5.5th value. This is interpreted as meaning halfway between the 5th and 6th values. In practice, if n is reasonably large the $+1$ term makes little difference; if $n = 100$, for example, we could take the median as the 50th value without much inaccuracy.

For a grouped frequency table, the simplest way to find the median is from the ogive. Go up the vertical axis to the $(n + 1)/2$ point, then across to the graph, and hence read off the median value on the horizontal axis. If you do this with Figure 7.9 (the ogive for the 'length of visit' data), you should find that the median is about 20 minutes. Perhaps a better way of expressing this – certainly if you are talking to someone who is not expert in statistics – is to say 'half the customers spent less than 20 minutes in the supermarket'. This demonstrates how the median, as well as being resistant to extreme values in the data, is easy to understand.

If you do not want to go to the trouble of plotting an ogive, the median can be calculated by a straightforward proportion argument. Recall the cumulative table for the 'length of visit' data:

FIGURE 7.10
Calculating the median

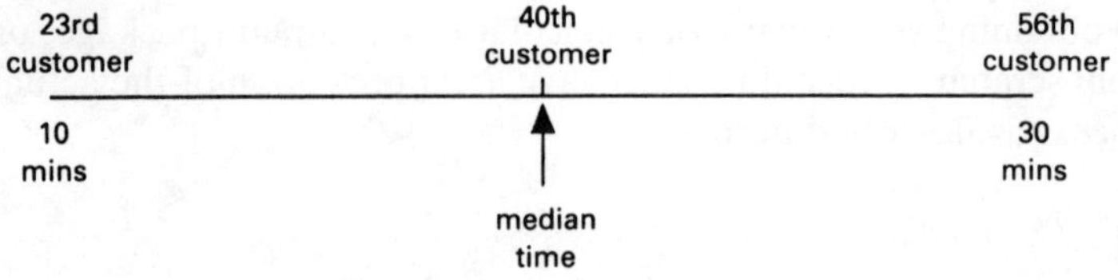

```
  23rd              40th                56th
customer         customer             customer
──────────────────────┬──────────────────────
  10                  ▲                  30
 mins                 │                 mins
                    median
                     time
```

Length of visit (mins)	No. of respondents (cumulative frequency)
< 10	23
< 30	56
< 60	80

Now the median is going to be at the 40th value (strictly the 40.5th, but we shall not worry about that). This clearly does not come in the 'under 10' group, where there are only 23 values. It must be in the next '< 30' group, since by the end of that group we have accumulated 56 values. The time belonging to customer number 40 will therefore be part-way between 10 and 30 minutes; and since customer number 40 is 17 people into this class, which has a total width of 33 people, we can say that the median is a fraction 17/33 of the way through the class, starting at 10 minutes.

Putting all this together, we find:

median = 10 minutes + (17/33) × 20 minutes
       = 20.3 minutes,

agreeing with the rough value we got from the ogive. If you find this hard to follow, Figure 7.10 may help to make it clearer.

(c) The mode

If you look at a frequency table, such as the one below for number of visits to XYZ Supermarket in the last month, one of the things you probably notice immediately is that the 'peak' frequency corresponds to the 3-visit class.

No. of visits to XYZ supermarket in the last month	No. of respondents
1	15
2	19
3	22
4	20
5 or more	4

So if you were asked to say how many visits a 'typical' shopper makes, this might well be your answer. The most common value in a set of data is called the *mode*, and is easily found in the case of an ungrouped frequency table such as we have here. However, with a grouped table the best we can do is to name the modal class. Moreover, there may be more than one mode, or indeed none at all (as with our distribution of men's weights, where every value occurs just once). So, on the whole, the mode, being rather ill-defined, is not a very useful measure of location.

7.4.3 *Measuring dispersion*

Each of the measures of location we have looked at has its corresponding measure of dispersion; we will look at them in the same order.

(a) The standard deviation

If you have learned some elementary statistics before, you will certainly have covered the standard deviation – but do you have a clear idea of what it *means*, as distinct from how it is calculated? The standard deviation is one of the most important and widely-used concepts in statistics, but it is not easy to grasp. We will try to give a justification for its method of calculation by using our sample of men's weights – with the sumo wrestler expelled and the 120-lb person reinstated!

The values in question are then 120, 131, 147, 152 and 160 lb, leading to the mean of 142 lb we calculated earlier. Now if we want an idea of how dispersed the individual values are, one reasonable question to ask is, 'How far are they from the mean?' If we denote the individual values by x as usual, then the mean will be $\bar{x}$, and the distances of the xs from the mean can be found by taking $x - \bar{x}$. The calculation is best set out in a table:

x	$x - \bar{x}$
120	−22
131	−11
147	5
152	10
160	18

Some of the $x - \bar{x}$ values – called the *deviations* of the xs from the mean – are positive, showing that x is bigger than the mean; those for xs lower than the mean are negative. The result of this is that if we add up all the deviations, to get an idea of the overall dispersion, we find that the total is zero. This will indeed always be the case; it is inherent in the way the mean is calculated. So if we want to devise a measure of dispersion based on the deviations, we have to eliminate the negative values in some way.

One possibility is simply to ignore them, and you may have come across a measure – the *mean absolute deviation* – based on this idea. However, ignoring information is not a very satisfactory procedure; a better way is to square each of the deviations. This gets rid of the minus signs (remember 'minus times minus = plus' from Chapter 1?) while preserving the information about large and small deviations. We thus get an extra column in our table:

x	$x - \bar{x}$	$(x - \bar{x})^2$
120	−22	484
131	−11	121
147	5	25
152	10	100
160	18	324
Total		1054

So the average squared deviation is 1054/5, or 210.8. We still have not quite finished; there is a bit of a problem with the units of this quantity, since all our original measurements were in pounds but by squaring the deviations we have ended up with a measure in square pounds! To return to our original units, we take the square root of 210.8, finally obtaining a figure of 14.5 lb which is the standard deviation of the weights.

Let us recap what we did to obtain this figure: we took the square root of the average of the squared deviations from the mean. We can express this symbolically as:

$$\text{standard deviation} = \sqrt{\frac{\sum (x - \bar{x})^2}{n}}$$

where n denotes the number of items in our set of data.

There is just one further small complication that we need to think about. Remember what we said earlier about the importance of being able to estimate measurements of a population from measurements on a sample? Well, the standard deviation as calculated from the formula above gives a biased estimate of the population standard deviation – it is a bit too small. To compensate for this, it can be shown mathematically that it is better to divide by $n - 1$ when you are only using sample data. This explains why, if you have a statistical calculator, it probably has two standard deviation keys, one labelled σ_n and one σ_{n-1}. The one with n should only be used if you are sure you have data about the whole population – and since in practice this is very unusual, you will find the n-1 divisor is nearly always used. Having said all this, when n is bigger than about 30 it makes very little difference to the answer!

The formula we have just derived is fine as a *definition* of standard deviation, but as a way of calculating it is a bit clumsy. It is easier to expand the bracket $(x - \bar{x})^2$ (you can work it out if you would like some practice at algebra) to get:

$$\text{standard deviation} = \sqrt{\frac{\sum x^2}{n} - \bar{x}^2}$$

And if we have a frequency table, each x occurs not just once, but f times, so we make a further modification:

$$\text{standard deviation} = \sqrt{\frac{\sum x^2}{\sum f} - \bar{x}^2}$$

The Greek letter σ (a small sigma, not to be confused with the capital sigma we use for 'add up') is generally used to represent the population standard deviation, and an ordinary s is used for the sample value.

Let us wind up our discussion of standard deviation by calculating it for the 'number of items' data for which we have already found the mean to be 14.1875. Here is the table we drew up earlier, to which a column of fx^2 values has been added – make sure you follow the calculation. The easiest way to find fx^2 is to multiply the fx by another x – your recollection of Chapter 2 should tell you that fx multiplied by x is fx^2.

No. of items purchased on this visit	No. of respondents f	x	fx	fx^2
1–5	6	3	18	54
6–10	15	8	120	960
11–15	37	13	481	6253
16–20	12	18	216	3888
21–30	7	25.5	178.5	4551.75
31 or more (50)	3	40.5	121.5	4920.75
Total	80			19667.5

Thus standard deviation = $\sqrt{(19667.5/80 - 14.1875^2)} = \sqrt{44.56}$ = 6.68 items.

You may feel not very much the wiser about what useful information this figure conveys. It is perhaps easiest to get a feel for the meaning of standard deviation by thinking of it in a comparative way – if shoppers at another supermarket had a mean of 12.2 items with a standard deviation of 4.1 items, that would suggest not only that the average number of items they buy is lower, but also that there is less variability in the number of items bought.

Another helpful fact is based on the theory of the *Normal Distribution* which you may study in your statistics or quantitative methods course: as long as the distribution is not too lopsided, all its values will lie within about three standard deviations from the mean. So in the present example, we could expect the number of items bought by all shoppers to range between 14.1875 − 3 × 6.68 and 14.1875 + 3 × 6.68, or from about 0 to 34 items. (The calculation actually gives a negative lower limit – that is because this particular distribution is a little asymmetric.)

We shall finish our discussion of standard deviation at this point; when you come to study it again, you should have an advantage in at least feeling comfortable with the calculations, and probably in having some idea of what it really means.

(b) The quartiles

Just as we divided a set of data into halves using the median, so we can divide it into quarters by values known as the *quartiles*: the lower quartile cuts off the lowest 25% of the data, and the upper quartile the highest 25%. They can be found in the same way as the median – either by reading off from the ogive, or directly from the cumulative frequency table. To illustrate, we will find the median and quartiles for the 'number of visits' data from the XYZ Supermarket survey.

No. of visits to XYZ supermarket in the last month	No. of respondents	Cumulative frequency
1	15	15
2	19	34
3	22	56
4	20	76
5 or more	4	80

The median will be the 40th value, which is 3 visits – we do not need to do the proportion calculation here because the table is not grouped. The lower quartile is the 20th value, which will be 2, and the upper quartile is the 60th value – that is 4. So we could say '25% of customers make 4 or more visits per month' and so on. The quartiles, by ignoring the most extreme 25% of values at each end, give us a range for the more 'typical' central 50%; they also avoid the need to close the open class '5 or more' at the top.

The median and quartiles, taken together, give useful information about the shape of a distribution, as well as its location and dispersion. For example, if you calculate their values for the 'length of visit' data from the XYZ survey (closing the top class at 60 minutes), you should find that the quartiles are about 9 and 35 minutes (we already calculated the median as roughly 20 minutes). So the lower quartile is a bit nearer to the median than the upper quartile, indicating that there is a long 'tail' on the upper end of the distribution.

(c) The range

As with the mode for measuring location, the most obvious measure of spread is the least useful. The *range* is just the distance from the smallest to the largest value in the data – so we could make statements like 'the wages of these workers range from 220 to 400 per week' or (less usefully) 'the range of the wages is £180'.

However, we cannot actually state a range for any of our sets of XYZ Supermarket data, because they all involve open classes. This is one clear drawback to the range. Another is the way it can be distorted by the presence of just one 'rogue' value, since it takes no account of any data values other than the two most extreme.

7.4.4 *Choosing a measure*

Different measures of location and dispersion have been devised because no one pair of measures is ideal in every circumstance. You need to make an informed choice of which to use for any particular set of data, and while it is not possible to lay down 'rules' for making this choice, we offer some fairly general guidelines:

1. If your data is reasonably symmetrically distributed, and/or you want to draw conclusions about population values based on your sample – use the mean and standard deviation.
2. If your data is markedly unsymmetric – use the median and quartiles.
3. If you just want a quick idea of the location and dispersion without any further calculation – use the mode and range.

Exercises 7.3

1. Find the mean and standard deviation of the following two samples of data:

4, 7, 9, 10, 15;
40, 70, 90, 100, 150.
What do you notice?
2. Find the mean, standard deviation, median and quartiles for the student mark data in Exercises 7.1, Question 1. Comment briefly on what the results tell you.

Appendix 1 Some Uses of Integration

The Definite Integral

As we saw in Chapter 5, integration enables us to determine a function from its derivative (the function giving its rate of change). Thus in practical situations where the rate of change of a quantity is known we can use integration to determine total 'accumulated' quantity.

For convenience in this Appendix, we will use the notation $F(x)$ to represent $\int f(x)\, dx$ **without** the constant of integration. So if $f(x) = x^4$, then $F(x) = x^5/5$, and so on. We will call $F(x)$ the *antiderivative* of $f(x)$, to distinguish it from the indefinite integral which includes the constant.

An example will now illustrate how integration enables us to determine total accumulated quantity from its rate of accumulation. In Question 3 of Exercises 5.2 you encountered the situation where the distance covered by a car was given as a function of the time that it had been travelling. Now imagine that we are given information, not about distance, but about *speed* as a function of time. Suppose the car travels at a steady 30 miles per hour. Then its speed v as a function of time can be written:

$$v(t) = 30. \tag{A1.2}$$

Clearly, when the car has been travelling for t hours the distance it will have covered is given by $30t$. Note that the expression $30t$ is virtually the same as the integral of the speed of the car with respect to time: $\int v(t)\, dt = \int 30\, dt = 30t + K$. The only difference between the two is the constant of integration K.

The apparent link between the integral of the speed and the distance travelled by the car is not coincidental. It rests on the fact that speed is the rate of change of distance with time. If $s(t)$ is the function giving the distance travelled by the car in t hours, then the rate of change of $s(t)$ will be $ds(t)/dt$. Therefore, when we say that the car travels at a steady 30 miles per hour what we mean is that $ds(t)/dt = 30$. It therefore follows from our definition of integration as inverse differentiation that $s = \int 30\, dt = 30t + K$.

We still have the constant of integration K to contend with. How can we use the integral to get the precise value of the distance travelled by the car in a given time period? Common sense tells us that, for example, in any period of 2 hours at a steady 30 miles per hour the car will travel exactly 60 miles – there is nothing indefinite about the result. So what happens to the K?

The answer is simple – in computing the distance travelled over any definite time-period, the K will cancel out. Suppose, for example, that we want to know the distance travelled in the half-hour period from $t = 1$ to $t = 1.5$. This will, of course, be 15 miles; but let us try to arrive at this result using the integral of the speed of the car. When $t = 1$ the distance travelled starting from $t = 0$ will be $s = \int 30 \, dt = 30t + K = 30 \times 1 + K$ miles. The K here gives the value of the distance when $t = 0$, so you can think of it as representing the initial distance of the car from some arbitrary origin. Likewise, when $t = 1.5$ the distance travelled from $t = 0$ will be $s = 45 + K$ miles. Thus the distance travelled from time $t = 1$ to time $t = 1.5$ is $[45 + K] - [30 + K] = 15$ miles, the K cancelling out as promised.

The arbitrary constant K will cancel in the same way for any definite time interval; the distance travelled from time t_1 to t_2 will be the value of the integral at t_2 minus that at t_1, and we do not need to bother about the constant.

So what we have found so far is that when a car travels at a constant 30 miles per hour the distance it travels between time $t = t_1$ and time $t = t_2$ is given by:

$$S(t_2) - S(t_1) \text{ where } S(t) = \int 30 \, dt.$$

What we are doing here is evaluating the difference in the values of the integral at t_1 and at t_2. Because this operation occurs so frequently, we have a special notation for it. We write:

$$S(t_2) - S(t_1) = \int_{t_1}^{t_2} 30 \, dt = [S(t)]_{t_1}^{t_2}. \tag{A1.3}$$

The expression $\int_{t_1}^{t_2} 30 \, dt$ is called the *definite integral* of 30 from t_1 to t_2, in contrast to the indefinite integral which we met in Chapter 5.

It can be shown – you can find the proof in many books on elementary calculus, but we are not going to go through it here – that these results for the distance travelled by the car actually hold good even if its speed is not constant. The only change is that $v(t)$, instead of being a constant 30, becomes a function varying with t. So in general, reverting to the more familiar x rather than t, we can say that:

- If $f(x)$ is the rate at which a quantity is accumulating then the total quantity accumulated over the interval x_1 to x_2 will be $F(x_2) - F(x_1) = [F(x)]_{x_1}^{x_2} = \int_{x_1}^{x_2} f(x) \, dx$, where $F(x)$ is the anti-derivative of $f(x)$.

The expression $\int_{x_1}^{x_2} f(x) \, dx$ is called the *definite integral of $f(x)$ with respect to x from $x = x_1$ to $x = x_2$*.

We can summarise the discussion in this section as follows:

1. If the indefinite integral of $f(x)$ with respect to x is $F(x) + K$, then the definite integral $\int_{x_1}^{x_2} f(x) \, dx = [F(x)]_{x_1}^{x_2} = F(x_2) - F(x_1)$.
2. This definite integral represents the increase in $F(x)$ as x increases from x_1 to x_2.

Two further examples will draw all these threads together.

EXAMPLE 1

What is the distance travelled by a car between $t = 1$ and $t = 2$ hour if its speed is $v(t) = 96t - 32t^2$ miles per hour?

SOLUTION

Since $v(t) = ds/dt = 96t - 32t^2$, the required distance is given by $\int_1^2 (96t - 32t^2) \, dt = [48t^2 - 32t^3/3]_1^2 = (48 \times 4 - 32 \times 8/3) - (48 \times 1 - 32 \times 1/3) = 69.33$ miles.

EXAMPLE 2

Find the definite integral of $1/x^2$ between $x = 4$ and $x = 5$.

SOLUTION

We want $\int_4^5 1/x^2 \, dx = [-1/x]_4^5 = (-1/5) - (-1/4) = 1/20$ or 0.05.

Notice that it is important to make sure that the function you are trying to integrate is well-behaved over the range of your definite integral. In the last example, for instance, we would have had problems if our range of integration had been from -1 to 1, since clearly something odd happens to $1/x^2$ at $x = 0$. Some integrals which involve problems of this kind can still be evaluated, but only by using more advanced methods which we are not going to consider here.

The integral as area

So far, we have defined integration as a mathematical process, and have seen how it can be used to compute the total quantity accumulated when its rate of accumulation is known. However, when we discussed differentiation, we were able to get more insight into the meaning of the process by giving the derivative a geometric interpretation, as the slope of a curve. It is therefore natural to ask whether a similar geometric interpretation exists for the integral.

It is easiest to answer this question by considering a specific example, the car travelling at a steady 30 miles per hour which we discussed earlier.

FIGURE A1.1
The definite integral as an area

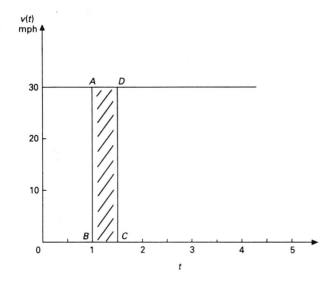

Let us now plot the speed of the car $v(t) = 30$ against t. The graph is, of course, a horizontal line (see Figure A1.1), since v does not change with t. If we want the distance travelled over a period of time, for example from $t = 1$ to $t = 1.5$, we can calculate it as distance = speed × time = $30 \times (1.5 - 1) = 30 \times 0.5 = 15$ miles. But 30×0.5 also gives us the area ABCD.
You should be able to see in the same way that whatever time interval t_1 to t_2 we might have chosen, the distance travelled would have been $30(t_2 - t_1)$ miles, which is the area between the graph $v(t)$ and the t-axis from $t = t_1$ to $t = t_2$.
So we have now got two ways of expressing the distance travelled in the time from t_1 to t_2: it is given by equation (A1.3), and also by the area between the graph of $v(t) = 30$ and the t-axis from $t = t_1$ to $t = t_2$. We can combine all this by saying that:

When a car travels at a constant speed $v(t) = 30$,
distance travelled in time t_1 to t_2
$= \int_{t_1}^{t_2} 30 \, dt$
= area between the graph of $v(t) = 30$ and the t-axis from $t = t_1$ to $t = t_2$.

In fact, it can be shown – though once again we are not going to go through the details – that this result is true even when the speed of the car is not a constant, but a function $v(t)$ which varies with t. Thus, in general,

- **When a car travels at a speed $v(t)$, distance travelled in time t_1 to t_2**
  $= \int_{t_1}^{t_2} v(t)\, dt$
  = area between the graph of $v(t)$ and the t-axis from $t = t_1$ to $t = t_2$.

There is, of course, nothing in this argument which is peculiar to problems involving speeds and distances; we would get exactly the same results for any function. So using x rather than t as our variable, we can say that:

- **The definite integral of $f(x)$ between $x = x_1$ and $x = x_2$ gives the area under the graph of $f(x)$ between the values $x = x_1$ and $x = x_2$.**

This gives us our geometric interpretation of the integral – it is an *area*. The result is useful in enabling us to compute peculiar-shaped areas under various curves, as the following example will show.

EXAMPLE

Find the area under the curve $y = 1/x$ between $x = 1$ and $x = 2$.

SOLUTION

This area will be equal to $\int_1^2 (1/x)\, dx = [-1/x^2]_1^2 = (-1/4) - (-1/1) =$ 3/4 square units.

The interpretation of the definite integral as an area enables us to derive a useful result: if $a < c < b$, then:

$$\int_a^b f(x)\, dx = \int_a^c f(x)\, dx + \int_c^b f(x)\, dx.$$

To see this, note that ABFG in Figure A1.2 is given by the definite integral $\int_a^b f(x)\, dx$. However, this area can be split into two smaller areas ABCD and DCFG. The definite integral $\int_a^c f(x)\, dx$ gives the area ABCD and $\int_c^b f(x)\, dx$ gives the area DCFG. But the sum of the two areas gives the area ABFG. Thus $\int_a^b f(x)\, dx = \int_a^c f(x)\, dx + \int_c^b f(x)\, dx$.

So far, our examples have all involved finding areas under curves which are positive over the range of integration. But now consider the case where the function $f(x)$ takes negative values in the interval $[a,b]$, so that the area is below the x-axis. See, for example, Figure A1.3 which shows the graph of $y = 4 - x^2$ for x from 0 to 4. Suppose we wish to find the area between the curve and the x-axis between $x = 2$ and $x = 4$.

Let us apply the definition of the area as integral, and see what happens. We have:

area ABC $= \int_2^4 (4 - x^2)dx = [4x - x^3/3]_2^4$
$= (16 - 64/3) - (8 - 8/3) = -10.66.$

FIGURE A1.2
Summing definite integrals

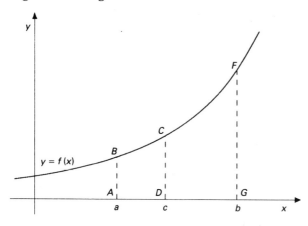

So we end up with a 'negative area'. However, in geometric terms the sign is of no significance, and we could say that the area ABC = 10.66 square units, ignoring the minus sign.

More tricky is the case when the range of integration includes areas both above and below the x-axis, as it would, for instance, if in Figure A1.3 we wished to find area EFCB between $x = 0$ and $x = 4$. If we try to do so in a simple-minded way, by evaluating the integral from $x = 0$ to $x = 4$ directly, you can verify that we get an answer of -5.33. However, if we work out the area in two sections, EFA plus ABC, then we find (you should check this!) that EFA has an area of 5.33, while we have already found that ABC has area 10.66. So the total area should be $5.33 + 10.66 = 16$ square units. What has gone wrong?

The answer is that if we work out the area by integrating over the whole range $x = 0$ to $x = 4$, then the part below the axis yields a negative contribution (as we saw above). So the value of $\int_0^4 (4 - x^2)dx$ is actually the *difference* EFA − ABC = $5.33 - 10.66 = -5.33$. The moral of all this is that you should always sketch a graph of the area you are trying to find; if this reveals that part of your function $f(x)$ is above and part below the x-axis, then to obtain a correct result it is essential to work out the areas of the two portions separately, and then add them together.

A business application

Much of our work on integration may have seemed rather theoretical, so we will end the discussion with an extended practical example.

FIGURE A1.3
Areas below the x-axis must be computed separately

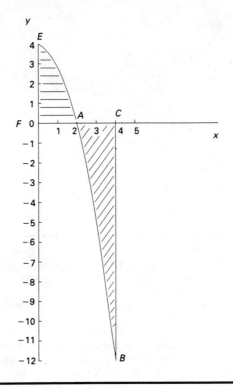

EXAMPLE

An oil company estimates that as an oil well is being developed, the costs of extraction of oil will rise over time, while revenue will fall as the well is progressively depleted. Annual costs are estimated to be $C(t) = 4 + t^2$, while annual revenue is estimated to be $R(t) = 16 - t$ where t is the age of the well in years. (Both $C(t)$ and $R(t)$ are in £m.)

(a) What is the total cost of running the oil well for the first two years?
(b) What is the total revenue from running the oil well for the first two years? What is the total profit over the same period?
(c) You should be able to see that the maximum total profit from the well will be obtained if it ceases production when annual cost

FIGURE A1.4
A business application of definite integrals

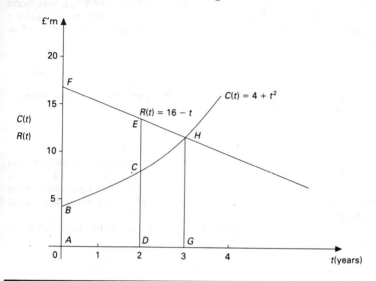

begins to overtake annual revenue. When should production from the well cease? What will total profit over the life of the well have been if production ceases at that time?

SOLUTION

(a) Figure A1.4 shows the annual cost and revenue as functions of time. Since both are varying continuously with time, we cannot simply add up the annual cost/revenue at $t = 1$, $t = 2$ and so on; instead, we must use the definition of the definite integral to find the total cost/revenue over a period of time. The total cost of running the well in the first two years will be represented by the area ABCD.

Area ABCD = $\int_0^2 (4 + t^2) \, dt = [4t + t^3/3]_0^2 = 8 + 8/3$
= £10.66m.

(b) The revenue in the first two years is given by the area ADEF.

Area ADEF = $\int_0^2 (16 - t) \, dt = [16t - t^2/2]_0^2 = 32 - 2$
= £30m.

The total profit over the first two years of the life of the well is 30 − 10.66 = £19.34m. This is incidentally the area BCEF.

(c) Annual cost is $4 + t^2$, and annual revenue is $16 - t$. We can read from Figure A1.4 the point where the graphs of these two functions intersect as $t = 3$. Alternatively, this value of t can be calculated algebraically. Annual cost and revenue are equal when $4 + t^2 = 16 - t$, or

$$t^2 + t - 12 = 0. \tag{A1.4}$$

This is a quadratic equation, which can be solved using the formula given in Chapter 4, to obtain the solution:

$$t = (-1 \pm \sqrt{(1 + 48)})/2.$$

This gives $t = 3$ or -4. However, we clearly cannot have the well operating for a negative period of time, so the root $t = -4$ is of no practical significance. We therefore conclude that the well has made its maximum profit after 3 years, and this confirms the value found from the graphs in Figure A1.4.

The cost of running the well for 3 years will be the area ABHG in Figure A1.4:

$$\text{Area ABHG} = \int_0^3 (4 + t^2) \, dt = [4t + t^3/3]_0^3 = £21\text{m}.$$

The revenue during the first three years is given by the area AGHF:

$$\text{Area AGHF} = \int_0^3 (16 - t) \, dt = [16t - t^2/2]_0^3 = 48 - 4.5 = £43.5\text{m}.$$

Thus the total profit over the life of the well is 43.5 − 21 = £22.5m. This is, of course, the area BHF.

Exercises A1

1. Determine the areas between the x-axis, the curves and the vertical lines at the x-values indicated in each of the following cases:

   (a) $f(x) = 2x + 1$, $x = 0$ and $x = 3$.
   (b) $f(x) = x^2 - 2x$, $x = 2$ and $x = 4$.
   (c) $f(x) = x^2 - 3x + 2$, $x = 0$ and $x = 3$.

2. A company has bought a new press costing £0.5m. It estimates that the savings as a result of using the press will be $S(t) = 0.5 - 0.05\,t$, where t is in years and $S(t)$ in £m per annum. After how many years will the press have paid for itself?

3. A radio manufacturer knows from past experience that her workers take less and less time per radio of a given model produced as more and more radios of that model are produced. That is, with experience workers become more productive. The curve giving time per unit of production as a function of units produced is commonly known as the 'learning curve'. The manufacturer has estimated that her workers' learning curve is $f(x) = 2 x^{-0.1}$, where x is the number of radios of a given model produced and $f(x)$ is work hours per radio. She bids to supply the first 2000 of her new model to a supermarket chain. What length of time should she estimate for the production of these radios?

4. A company launches a new product for which the marginal cost of a unit is £0.5 and the fixed cost of launching the product is £100 000. The company estimates that as more and more units are sold, marginal revenue (that is, revenue per additional unit) will rise but this rise will gradually become negligible due to an increasing number of competitors entering the market. More precisely, it estimates that marginal revenue per unit is $MR(x) = 0.4 + x^{-0.125}$, where x is number of units sold and $MR(x)$ is in £s. When should the company discontinue production of the unit in order to maximise profit from its sales? What will total profit over the life of the product be?

Appendix 2   Glossary of Technical Terms

Absolute maximum (minimum)   The greatest (lowest) value attained by a function over its entire domain. Also sometimes called a *global* maximum (minimum).

Absolute value   The value of a number when any negative sign is ignored. For example, the absolute value of -3 is 3; the absolute value of $+8$ is 8. Also called *modulus*, and generally written as $|-3|$, etc.

Asymptote   A line which a graph approaches, while never actually meeting it. An example would be the graph of $y = 1/x$, which has both the x and y-axes as asymptotes.

Bar chart   A means of displaying frequency data using bars of varying heights to represent the frequency of occurrence in different categories of data.

Base   The base of a power is the number being raised to the power; so the base in the expression y^4 is y. The base of a logarithm is similarly defined.

Coordinates   The values relative to horizontal and vertical axes which enable us to fix the position of a point on a graph.

Denominator   The number or expression on the bottom of a fraction; for example, the denominator of 3/5 is 5; the denominator of $x/(x + 6)$ is $x + 6$.

Dependent variable   In an expression such as $y = 3x^2 - 2x + 1$, y is known as the dependent variable because its value depends on the value we give to x. The dependent variable is generally plotted on the vertical axis of a graph.

Derivative   The result of performing a differentiation. The derivative of x^4 with respect to x is $4x^3$.

Difference   The result of subtracting one number from another. For example, the difference of 7 and 5 is 2; the difference of 5 and 7 is -2.

Discrete data   Data which can take only certain specified values. An example would be the number of broken eggs in a box, which must be 0, 1, 2,. . . .

Dividend   The number being divided in a division calculation. For example, the dividend in the calculation $5 \div 2$ is 5.

Divisor   The number by which we are dividing in a division calculation. For example, the divisor in the calculation $5 \div 2$ is 2.

Domain (of a function) The range of values of the independent variable(s) over which the function is defined. For example, if we are told that for production levels n between 100 and 1000 units, variable costs are £3 per unit, and fixed costs £400, then the total cost function $C = 400 + 3n$ will have domain $100 \leq n \leq 1000$.

e The so-called 'base of natural logarithms'; a number with the approximate value 2.71828, which occurs naturally in various mathematical processes, for example as the sum of the series $1/0! + 1/1! + 1/2! + \ldots$.

Exponent The power in an expression such as x^3; here the exponent of x is 3.

Exponential function The function e^x, where e is the base of natural logarithms. $y = e^x$ is the only function with the property that $dy/dx = y$.

Exponential growth Growth according to a power law, such as $y = 10^x$ or $y = 1.1^t$. An example would be compound interest. The exponential function $y = e^x$ is a special case of exponential growth.

Factorial The factorial of a positive integer is the result of multiplying that integer by all positive integers smaller than itself. The factorial of n is written $n!$. Thus $5! = 5 \times 4 \times 3 \times 2 \times 1$. $0!$ is defined to be 1.

Frequency In statistics, the number of times a particular value or range of values occurs in a set of data. For example, in the set of data 1, 1, 2, 2, 2, 3, 3, 4, the frequency of 2 would be 3.

Frequency distribution A table showing the frequency of each value or range of values within a dataset. Also called a frequency table. If ranges are used, we refer to a grouped frequency distribution.

Function A relationship between a dependent variable and one or more independent variables; often represented by an equation. For example, the general linear function of a single independent variable can be written as $y = ax + b$.

Gradient The slope of a graph. If the graph is plotted on (x, y)-axes then the slope is the increase in y per unit increase in x. A positive slope means that the graph runs uphill from left to right; a negative slope (i.e. y decreasing as x increases) means the graph runs downhill from left to right.

Histogram A way of displaying a grouped frequency distribution by means of blocks, whose width is proportional to the width of the classes in the distribution, and whose area corresponds to the frequency of the class.

Hyperbola The shape of graph which results from plotting a function containing a term of the form $1/x$, $1/x^2$, etc. For examples see page 94.

Inequality An algebraic or arithmetical expression representing the relationship between the sizes of two quantities. The signs $<, >, \leq, \geq$ are used to represent 'less than', 'greater than', 'less than or equal to', and 'greater than or equal to'. So $2 < 6$, etc.

Integer 'Whole number', such as 11, 362 or -50.

Integration The process of finding a function given its derivative function. Represented by the symbol $\int$.

Intercept The point where a graph crosses the vertical axis.

Linear equation An equation in which no variable is raised to any power other than 1 or multiplied by any other variable. A linear equation in a single variable can always be put into the form $ax + b = 0$.

Linear function A function in which the dependent variables only occur raised to the power 1. Can be expressed in the form $y = a_0 + a_1 x_1 + a_2 x_2 + \ldots$

Logarithm The logarithm of x to the base y is z, where $x = y^z$. So, for example, the logarithm of 243 to the base 3 is 5, because $243 = 3^5$. *Common* logarithms are those to base 10; *natural* logarithms have base e.

Logarithmic function A function of the form $y = \log x$.

Lowest common denominator The smallest number into which the denominators of a group of fractions will all divide. For example, the lowest common denominator for the fractions 1/4, 1/3 and 1/6 is 12, because this is the smallest number into which 4, 3 and 6 will all divide. Used in addition and subtraction of fractions.

Local maximum The function $y = f(x)$ has a local maximum when $x = x_0$ if $dy/dx = 0$ when $x = x_0$, and dy/dx is positive to the left of x_0 and negative to the right of it; or equivalently, if d^2y/dx^2 is negative at $x = x_0$. This means that $f(x_0)$ is greater than the value of $f(x)$ anywhere else in the immediate neighbourhood of x_0. A local maximum may also be an absolute maximum (see above).

Local minimum The function $y = f(x)$ has a local mimimum when $x = x_0$ if $dy/dx = 0$ when $x = x_0$, and dy/dx is negative to the left of x_0 and positive to the right of it; or equivalently, if d^2y/dx^2 is positive at $x = x_0$. This means that $f(x_0)$ is smaller than the value of $f(x)$ anywhere else in the immediate neighbourhood of x_0. A local minimum may also be an absolute minimum (see above)

Mean The arithmetical mean, which is usually contracted to 'mean', is what is colloquially called the 'average'; so the mean of 3, 4 and 5 is $(3 + 4 + 5)/3 = 4$, and so on.

Median The central value in an ordered set of numbers (or the mean of the two central values if there is an even number of values). For example, the median in the set 21, 33, 47, 68, 74, 91 is $(47 + 68)/2 = 57.5$.

Mode The most common value in a group of numbers. The mode in the group 1, 1, 2, 2, 2, 3, 3, 4 is 2.

Modulus See Absolute value.

Natural logarithm See Logarithm.

Numerator The number or expression on the top of a fraction. For example, the numerator of 11/12 is 11; the numerator of $(x + 2)/(x - 3)$ is $x + 2$.

Ogive A graph plotted from a cumulative frequency table.

Origin The point (0, 0) on a graph, which is often, but not invariably, the point where the coordinate axes meet.

Parabola The shape of graph which results from plotting a quadratic function. For examples, see page 79.

Percentage A fraction expressed on a denominator of 100; the denominator is not explicitly written. Thus 25% means 25/100, etc.

Pie-chart A means of displaying frequency data using sectors of a circle to represent subgroups of the data.

Point of inflexion A point on a graph where $dy/dx = 0$, but dy/dx does not change sign.

Power An expression such as x^n is called the nth power of x. More loosely, we sometimes refer to n as the power of x in this expression.

Product The result of multiplying two or more numbers or expressions. So the product of $2x$ and x^2 is $2x^3$.

Quadrant One of the four parts into which the coordinate axes divide the Cartesian plane.

Quadratic A quadratic expression in x is one where the highest power of x occurring is a square. A quadratic equation is one involving only quadratic expressions; a quadratic equation in one variable can always be put into the form $ax^2 + bx + c = 0$.

Quartile A number which divides an ordered set of figures into quarters. The smallest quartile is the lower quartile, the middle one is the median (see above) and the biggest is the upper quartile.

Quotient The result of dividing two numbers or expressions. The quotient of 12 and 3 is 4; the quotient of 3 and 12 is 0.25.

Range The difference between the largest and smallest values in a set of data.

Rate of change If y is a function of x, the rate of change of y with x is the increase in y per unit increase in x. A negative rate of change means y decreases as x increases.

Root of an equation $x = a$ is a root of an equation in x if substituting $x = a$ into the equation makes the two sides equal. For example, $x = 2$ is a root of $2x + 8 = 10 + x$ because when $x = 2$, both sides of the equation are equal to 12. A root is said to *satisfy* the equation in question.

Scientific notation A compact form of notation in which numbers are written as a figure with one digit before the decimal point, multiplied by a power of ten. Thus 25 021 in scientific notation would be written 2.5021×10^4, or more usually as 2.5021E4. Similarly, 0.00329 would become 3.29×10^{-3} or 3.29E−3.

Significant figures The number of digits which actually carry information, as distinct from those which simply indicate place-value. Thus in 30 270, there are four significant figures: the 3, 2, 7, and the zero between the 3 and the 2. The last zero is, however, not significant, since it only indicates the overall place-value.

Slope The rate of change of the dependent variable in a graph per unit increase in the independent variable.

Solution The solution of an equation is the value or set of values of the unknown(s) which make the two sides of the equation equal.

Standard deviation A measure of the spread of a set of data, found by taking the square root of the mean of the squared deviations of the data values from their mean.

Sum The result of adding two or more numbers or expressions. The sum of 11, 2 and $3x$ is $13 + 3x$.

Summation sign (Σ) Σx means 'add up all the values of x'.

Tangent A line which meets a curve only at a single point.

Turning point A general term for a local maximum or minimum (see above).

Appendix 3 Solutions to Exercises

Chapter 1 Arithmetic Operations

Exercises 1.1

1. (a) −5; (b) −6; (c) −5; (d) (+)56 (the + is not usually written).
2. Sum is −3 (add).
3. Product is −72 (multiply).
4. Overall profit is £47 800 000.
5. 2800 pence, or £28.

Exercises 1.2

1. (a) $\frac{16}{15}$, or $1\frac{1}{15}$; (b) $\frac{5}{18}$;
   (c) $\frac{33}{84}$ or $\frac{11}{28}$; (d) $\frac{6}{75}$ or $\frac{2}{25}$.
2. Sum is $\frac{14}{9}$ or $1\frac{5}{9}$.
3. Product is $\frac{24}{36}$ or $\frac{2}{3}$.
4. 24.
5. $\frac{3}{24}$ or $\frac{1}{8}$ (the calculation is $\frac{1}{12} \times \frac{3}{2}$).
6. There are two ways of doing this calculation: *Either* the proportion paying by cash will be $1 - (\frac{2}{3} + \frac{1}{8}) = \frac{5}{24}$ (why do we need the brackets?) and $\frac{5}{24} \times 18\,000 = 3750$,
   OR $\frac{2}{3}$ of 18 000 = 12 000 credit-card payers;
   1/8 of 18 000 = 2250 cheque payers.

So the remaining 18 000 − (12 000 + 2250) = 3750 are cash payers.
7. Possible equally-weighted splits are:
   2 × 60;
   3 × 40;
   4 × 30;
   5 × 24;
   6 × 20;
   8 × 15;
   10 × 12;
   12 × 10;
   etc.
   i.e. 14 possibilities altogether – and, of course, in practice subjects can have different weights, so that 1 at 40 points and 4 at 20 are also feasible.

   In contrast, with 100 points the combinations are 2 × 50, 4 × 25, 5 × 20, 10 × 10, 20 × 5, etc. – total of only eight. Schemes based on 100 points would thus have far less flexibility.

8. $\frac{28}{1000} = \frac{7}{250}$, whereas $\frac{8}{1000} = \frac{2}{250}$. Thus the rate was $\frac{7}{2}$ times, or three-and-a-half times, greater in 1953.

Exercises 1.3

1. (a) 4.031; (b) −4.32; (c) 20;
   (d) 208; (e) 0.8355;
   (f) 1445.1666 . . . (the 6 recurs).
2. 0.8888 . . . (the 8 recurs).
3. 13/40 (325/1000 can be cancelled by 25).
4. The sum is 102.19 × 10.13 = 1035.1847 (in fact this would have to be rounded to 1035.18 – see next section – since the kroner only splits into 1/100ths).
5. The price is 13 492/1.43 = £9434.965.

Exercises 1.4

1. 8.927; 24.369; 0.005.
2. 10 200; 0.0306; 2.67.
3. £24 279 to the nearest pound could be anywhere between £24 278.50 and £24 279.49. Similarly, £18 730 to the nearest £10 could be between £18 725.00 and £18 734.99; and £19 350 to the nearest £50 could be between £19 325.00 and £19 374.99. So the total could lie anywhere in the range £62 328.50 to 62 389.47, or approximately £62 359.00 +/− £30.50. Thus it would be very rash to quote the figure simply as £62 359, implying that it is accurate to the nearest £1, when strictly the answer should be £62 359 with a possible error of £30.50 either way.

Exercises 1.5

1. 42%.
2. 36.
3. £19 756 (take 88% of the original bill).
4. £2.90 (take 116% of the original bill).
5. 0.02.
6. 22.5%.
7. 59.5% (to one decimal place).
8. 19.4% (to one decimal place – the actual decrease is 67, which as a percentage of last year's figure of 346 is 19.4%). This figure is not very useful unless we also know the total numbers of mortgage-holders involved; if (admittedly a rather extreme case) the society had 3460 customers last year, and only 558 this year, then last year's 346 defaulters constituted only 10% of the total, whereas this year's form 50% – a serious position totally obscured by simply saying 'number of defaulters fell by 19.4%'. So generally a fairer comparison could be made by stating what *percentage* of customers defaulted in each year.
9. 279 000 as a percentage of 1 087 000 is 25.7%, so the number has fallen by $100 - 25.7 = 74.3\%$ of the 1953 figure.
10. Remember that you *cannot* simply say the rate over 10 years is $10 \times 3\% = 30\%$. Instead, using the hypothetical population of 10 000 as a base, you need to say population after 1 year = $10\,000 \times 103\%$ = 10 300; after two years it is $10\,300 \times 103\% = 10\,609$, and if you carry on like this for the whole ten years, you should find the final figure is 13 439 – an increase of 34.39%. (*Hint*: the constant multiplier facility which most modern calculators possess can make this kind of computation much easier.)

Exercises 1.6

1. 75 (add $3 + 9 + 15 + 21 + 27$).
2. (a) 3; (b) 8; (c) 4.
3. (a) true; (b) true; (c) true; (d) false.
4. y could take values of $-1, 0, 1, 2$ or 3 (the list does not include -2 or 4 because the inequality signs do not include $(=)$.
5. Selling price $\leq 1.5 \times$ cost.
6. (a) 720; (b) 30 240; (c) 24.

Chapter 2 Algebraic Expressions

Exercises 2.1

1. (a) 196; (b) 25; (c) 1296; (d) 5; (e) -7; (f) no real root.
2. (a) $(a + 8)$; (b) $(a - 2)$; (c) $-2x$.

Exercises 2.2

1. (a) $-15\,625$; (b) ± 25; (c) 1; (d) $1/5 = 0.2$; (e) $\sqrt{125} = \pm 11.18$; (f) $25/4 = 6.25 = (2.5)^2$.
2. (a) $4/25$; (b) 25; (c) 4; (d) $1/10 = 0.1$; (e) ± 2; (f) 1875; (g) $-1/8$.
3. (a) $(cb^2 + ad^4)^{1/2}/(bd^2)$; (b) $1/(1 + b)$; (c) b^2; (d) 0; (e) $(1 + a^2)^2/a^2(1 + a)^2$.

Exercises 2.3

1. (a) $2 \times 6 = 12$; (b) $-(-2) = 2$; (c) $a - 8 + 2b$;
 (d) $4b - 4a$; (e) $-2a^2 + ab = ab - 2a^2$; (f) $a^2 - b^2$;
 (g) $bc - b^2cd - b^3$;
 (h) $2a - 3a^2 - a^2b + a^2c - 2b + 3b^2 + ab^2 - abc$;
 (i) $4ab - 2ab^2c + 2abcd - 4cb + 2b^2c^2 - 2c^2bd$.
2. (a) Let W = no. of weeks timber will cover.
 Then $W = (I + N \times 30)/15$.
 (b) If $T \leq 3$ then $C = PT$. If $T > 3$ then $C = 3P + 0.85 P (T - 3)$.
 (c) Let B be the number of seats booked.
 If $B \leq 20$ then Cost in £s is Cost $= 0.9 \times F \times B$.
 If $20 \leq B \leq 120$ then Cost in £s is
 Cost $= (0.9 - 0.002 (B - 20)) F \times B$.
 If $B \geq 120$ the maximum 30% discount on F is reached. So
 Cost $= 0.7 \times F \times B$.

Exercises 2.4

(a) $2(a + 8b)$.
(b) $3a (a - 9b + 7ab^2)$.
(c) $2(a + b) - a(a + b) = (2 - a)(a + b)$.
(d) $a^2(a + b) - b^2 (a + b) = (a^2 - b^2) (a + b)$.
 Note also that $(a^2 - b^2) (a + b) = (a - b) (a + b) (a + b) = (a - b) (a + b)^2$.
(e) $(3c)^3 - 5^3 = (3c - 5) (9c^2 + 15c + 25)$.
(f) $[a^2 - 4(b + 2)^2][a^2 + 4(b + 2)^2]$.
(g) $(a + 1)(a^2 - a + 1)$.

Chapter 3 Functions and Graphs

Exercises 3.1

1. See Graph 1.
2. See Graph 2.

1. Graph of $y = 6 + 2x$, $0 \leq x \leq 5$

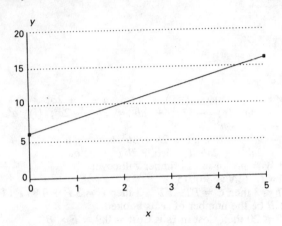

2. Graph of $y = 2 - 3x$, $-2 \leq x \leq 2$

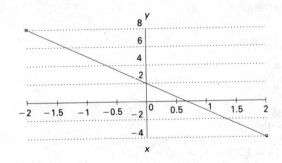

3. (i) Let V be the value of the car in £s after t years. Then
$V = P - 0.2\,Pt = P(1 - 0.2t)$.
   (ii) Let M be the mileage the car covered during the third year. Then the total cost of running the car during the third year, in £s, is $C = 0.2\,P + 0.2\,M$.

(iii) For the function $V = P - 0.2\,Pt$, a sensible domain for t would be that which keeps V non-negative. For this we need the domain $0 \leq t \leq 5$. (If t goes above 5, the value of the car is negative which would mean we would pay someone to take it. (Depending on its condition after 5 years this may well be the case!))

For the function $C = 0.2\,P + 0.2\,M$, a sensible domain for M would be whatever mileage we believe the car can possibly cover in a year without becoming a write-off and with running costs remaining at 20 pence per mile. (For large annual mileages there may well be increased maintenance costs.)

4.

	y-intercept	Slope	Rate of change
(a)	-2	2	2
(b)	6	3	3
(c)	$\frac{3}{2}$	-1	-1
(d)	15	10	10
(e)	1	$-\frac{1}{3}$	$-\frac{1}{3}$

The graph of (a) is not perpendicular or parallel to the graph of another function.

The graphs (b) and (e) are perpendicular.

The graph of (c) is not perpendicular or parallel to the graph of another function.

The graph of (d) is not perpendicular or parallel to the graph of another function.

5. Let n be the no. of leaflets printed and C the cost in £s.
If $n \leq 1000$ then $C = 100$. If $n \geq 1000$ then $C = 0.1n$.

6. Let PC be the printer's cost in £s for printing n leaflets.
Then $PC = an + b$.
a is the rate of change of PC with N.
We are given that when $n = 10\,000$, $PC = 500$, and when $n = 25\,000$ $PC = 900$. So an increase of $25\,000 - 10\,000 = 15\,000$ in n brings about an increase of £900 − £500 = £400 in PC. Hence the rate of change of PC with n is $a = 400/15\,000 = 4/150$, or about 2.7 pence per leaflet.

We can compute the value of b from the fact that when $n = 10\,000$, $PC = £500$. Hence, $500 = an + b = (4/150)\,10\,000 + b = 266.67 + b$. So b must be $500 - 266.67 = £233.33$. (The same value for b is obtained if $n = 25\,000$ and $PC = £900$ are used.)

The function linking the printer's cost to the number of leaflets printed is therefore:

$$PC = 4n/150 + 233.33, \text{ or } PC = 0.027\,n + 233.33.$$

7. The fixed cost is £233.366 and the additional cost of each leaflet is £4/150 = £0.0266 or 2.667 pence to three decimal places. To break even the printer needs $0.027\,n + 233.33 = 0.1\,n$ or $n = 3196$ leaflets for the run.

Exercises 3.2

1. (i) See graphs (a)–(d).

(a) Graph of $y = 3x^2 - 10$, $-4 \leq x \leq 4$

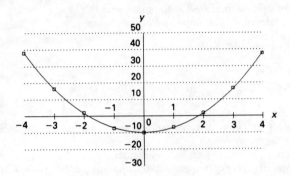

(b) Graph of $y = 10e^x$, $-4 \leq x \leq 4$

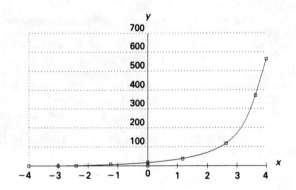

(c) Graph of $y = 100/(x - 1)$, $-4 \leq x \leq 4$

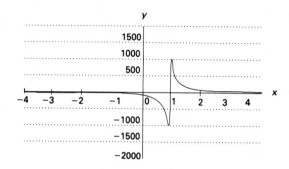

(d) Graph of $y = x^2 - x + 3$, $-4 \leq x \leq 4$

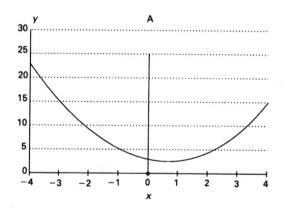

Types of curve: (a) parabola; (b) exponential; (c) hyperbola, asymptotic on $x = 1$ and the x-axis; (d) parabola.
(ii) See graph (e).

(e) Graph of $y = \log 100x$, $0 < x \leq 4$

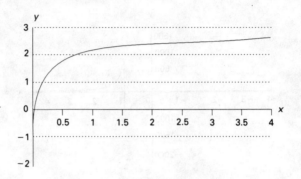

2. (i) Revenue £R = price per copy × copies sold.
   If price = £$(0.65 + P)$ then copies sold will be
   $(500 - 600P) \times 1000$. Hence $R = (500 - 600P) \times 1000(0.65 + P)$.
   (ii) See the graph below. The graph is a parabola.

Graph of $R = (500 - 600P) \times 1000(0.65 + P)$, $0 \leq P \leq 0.7$

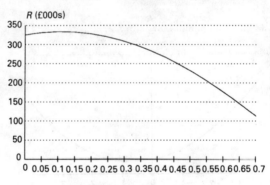

(iii) We saw in (i) that the revenue is
$R = (500 - 600P) \times 1000(0.65 + P)$. Note that the price per copy is £$(0.65 + P)$. So when the price per copy is 0.65, $P = 0$ and the revenue $R = £325\,000$. As the price rises above £0.65, P rises above 0. We can see from the graph in (ii) that initially revenue R rises as P rises but then revenue begins to fall. When the price per copy reaches just under £0.75 revenue R reaches a maximum of about £330 000. As the price per copy and P increase further, revenue falls. For example, when the price per copy is £1, P is 0.35 and so the revenue is £290 000. This decrease in revenue is caused by the lost demand for the paper when its price becomes too high.

3. (i) Let the tax allowance be £$y(t)$ after t years:
After 1 yr: $y(1) = 0.25\,P = P/4 = £5000$.
2 yrs: $y(2) = (P - P/4) \times 1/4 = 3P/16 = £3750$.
3 yrs: $y(3) = (P - P/4 - 3P/16) \times 1/4 = 9P/64 = £2812.50$.
4 yrs: $y(4) = (P - P/4 - 3P/16 - 9P/64) \times 1/4 = 27P/256 = £2109.38$.
5 yrs: $y(5) = (P - P/4 - 3P/16 - 9P/64 - 27P/256) \times 1/4 = 81P/1024 = £1582.03$.

(ii) See the graph below.
The graph is that of an exponential function.

(iii) When plotted on semi-log paper, the graph becomes a straight line, as can be seen from the graph (a) on page 228.
If you look back at part (i), you should be able to see that we could write $y(1) = 5000 \times 0.75^0$, $y(2) = 5000 \times 0.75^1$, $y(3) = 5000 \times 0.75^2$, and in general $y(t) = 5000 \times 0.75^{t-1}$. This formula makes explicit the fact that we are dealing with an exponential function. If we take logs of both sides of this equation, we get:

Graph of tax allowances against time

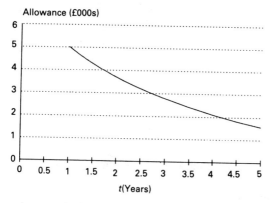

$$\log y = \log(5000 \times 0.75^{t-1}) = \log 5000 + (t-1)\log 0.75,$$

so

$$\log y = At + B,$$

where $A = \log 0.75$ and $B = \log 5000 - \log 0.75$

$= \log(5000/0.75) = \log(6666.67)$.

Now you can see why plotting $\log y$ against t gives a straight line. Moreover, the slope of the line is equal to the log of the annual reduction factor $= \log 0.75$. Because the log of a number less than 1 is

negative, this means that the line has a negative slope, as seen in the figure.
4. $\log a + (2/3) \log b + \log c - 2 \log a = (2/3) \log b + \log c - \log a$.
5. See graph (b) below. The graph is a hyperbola. It indicates that costs rise ever faster as the precision with which the life of tyres is estimated increases. (Note that the smaller d is, the higher the precision with which the life of the tyres is estimated.) The practical implication of this is that there will be a point beyond which the benefits of improving the precision of life estimates will be far outweighed by the costs.

(a) Tax allowances plotted on semi-log scale

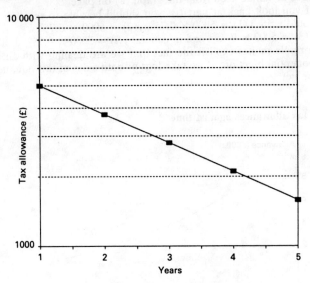

(b) Graph of costs against the accuracy of estimated life of tyres

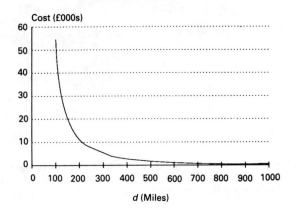

6. See graphs (a) and (b) below. Whereas the curve obtained on ordinary linear paper would be hard to identify as $y = 10^x$ simply by inspection, the semi-log graph is a straight line. This indicates that we are dealing with a constant percentage rate of change in y.

(a) $y = 10^x$ **(linear scale)**

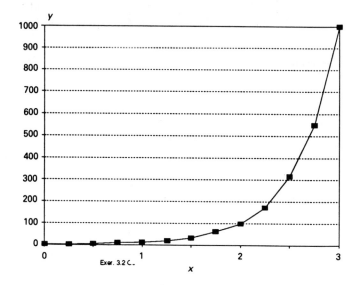

(b) $y = 10^x$ **(semi-log scale)**

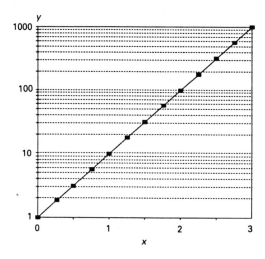

More precisely, the straight line shows that the relationship is of the form $\log y = ax$ (there is no '$+ b$' term in this straight line equation since the graph passes through the 'origin', where $x = 0$ and $y = 1$). When x increases by 1 unit, y increases by a factor of 10, as can be seen from the graph.

7. See the graph below. The graph shows that the rate of inflation was fairly constant over the period, with a slight tendency to decrease towards the end of the period. (Note that nothing could be said about the *rate* of inflation from examining a linear plot of the data.)

Graph showing the rate of inflation between 1987 and 1992

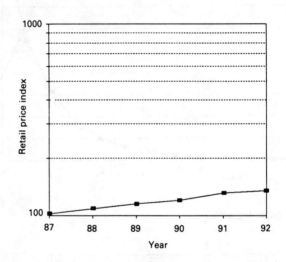

Chapter 4 Equations

Exercises 4.1

(a) $x = 2$; (b) $x = -1$; (c) $x = -35$; (d) $x = 3$.

Exercises 4.2

(a) $x = 3$ or -3;
(b) $x = 0$ or $3/2$;
(c) $x = (15 \pm \sqrt{209})/2$, $x = 14.73$, or $x = 0.27$;
(d) $x = 1/3$ or $-1/3$;
(e) $x = (2 \pm \sqrt{84})/10$, $x = 1.12$, or $x = -0.72$.

Exercises 4.3

1. (a) $x = (a + 5)/(1 - 2a)$
   (b) $x = 2/(3a - 4)$
   (c) $8(a + 1)/(2a - 1)$
2. Solving for x we have $x = (10y - 2)/35$.
   Hence $x = 0$, $y = 1/5$ is one combination. Another combination is obtained by setting $y = 0$. The resulting value of x is $x = -2/35$. There is an infinite number of combinations of x and y values which satisfy the equation, for example $(x = 1, y = 3.7)$, $(x = 2, y = 7.2)$ and so on.
3. Factorising we have $ax(5ax - 2) = 0$. Thus either $x = 0$ or $5ax - 2 = 0$ in which case $x = 2/5a$. This last value of x is only defined if $a \neq 0$.
4. Square rooting both sides we have $y^{1/2} = \pm 1/(1 - x)$. Hence:

   $$x = 1 \pm y^{-1/2} \text{ and so } x = 1 + y^{-1/2} \text{ or } x = 1 - y^{-1/2}.$$

5. Let w be the wife's, d the daughter's and s the son's amount in £million. Then $w + d + s = m$ and $w = 0.75d + 0.75s$. So we have $1.75d + 1.75s = m$. Hence, $s = (m - 1.75d)/1.75$.

Exercises 4.4

1. (a) $x = 16, y = 2$; (b) $x = 0, y = 3$;
   (c) $s = 4, t = 2$; (d) Equations are inconsistent – no solutions exist;
   (e) $a = b = 1/7$; (f) Equations are the same – an infinite number of solutions exists.
2. If the two numbers are x and y, then $x + y = 7$ and $x - y = 15$.
   Solving gives $x = 11, y = -4$.
3. Let s be the price of a single ticket, and r be the price of a return. Then the information in the question translates into the equations:

   $$2s + r = 3.3$$
   $$3s + 2r = 5.7,$$

   giving the price of a single as 90p, that of a return as £1.50.
4. (a) See the graph on page 232. The graph of the two functions (plotted on a single pair of axes) is shown on page 232.
   (b) We want to find where $C = R$, which means that:

   $$500 + 3N = 6N(24 - N).$$

   Multiplying out the brackets:

   $$500 + 3N = 144N - 6N^2,$$

   and collecting all terms on the LHS gives:

Graph of costs and revenue

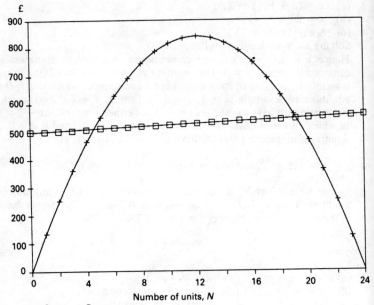

Key: □ Costs   + Revenue

$$6N^2 - 141N + 500 = 0.$$

You should recognise this as a quadratic equation, with $a = 6$, $b = -141$, and $c = 500$. The formula therefore gives:

$$N = \frac{141 \pm \sqrt{(141^2 - 4 \times 6 \times 500)}}{2 \times 6}$$

or

$$N = 4.35, \text{ or } 19.15.$$

You can see by examining the graph that these are the points where the two graphs intersect. Geometrically speaking, the points of intersection of the graphs are the simultaneous solution of the equations of the graphs.

From a practical point of view, $N = 4.35$ is the point at which a profit begins to be made (the *break-even point*), while if the firm continues production beyond $N = 10.15$, it will start to make a loss again, since costs exceed revenues.

Exercises 4.5

1. (a) $x \geq 16$; (b) $y \leq -7/2$; (c) $x \geq 24/5$; (d) $x \leq -2$.
2. Inequalities (a) – (c) respectively simplify to $x \geq 2, x \leq 2, x \geq 3/4$, and so only when $x = 2$ are all three inequalities satisfied.
3. Inequalities (a) – (c) respectively simplify to $x \geq 4(1 + a)/3, x \leq 5(1 + a)$, and $x \geq (1 + a)/4$. Hence only when $4(1 + a)/3 \leq x \leq 5(1 + a)$ are all three inequalities satisfied. (Note that $4(1 + a)/3 > (1 + a)/4$.)
4. The farmer's revenue will depend on the number of acres he can cultivate next season. He will cultivate the maximum number of acres his own time and the available combine harvester time will permit.

   Let A be the number of acres the farmer can cultivate next season. Then the limits on A are as follows:

   $A \leq 20$ Only 20 acres are available.
   $12.5\, A \leq 200$ Each acre requires 12.5 hours of the farmer's time and he can devote only 200 hours of his time to the field.
   $2\, A \leq 30$ Each acre requires 2 hours of combine harvester time and only 30 hours are available.

   We need to solve the three simultaneous inequalities to find the largest value A can take. The first inequality does not require any further manipulation. The second inequality gives:

   $A \leq 200/12.5$, or $A \leq 16$.

   The third inequality gives:

   $A \leq 30/2$, or $A \leq 15$.

   Plotting on the number line the ranges of A-values which satisfy the three inequalities we find the position in the graph below.

Graph of three simultaneous equations

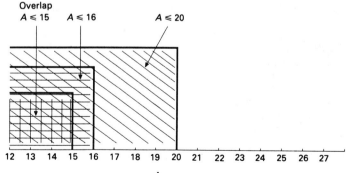

The range $A \leq 15$ is where the three ranges of possible values for A overlap.

Thus the range $A \leq 15$ satisfies all three inequalities. Hence the farmer can cultivate at most 15 acres of the field.

The revenue the farmer can expect from the 15 acres is:

$$\text{Revenue} = £500 \; x \; 15 = £7500.$$

5. Since the carpenter is committed to making 5 tables next week, each one taking five hours, $5 \times 5 = 25$ hours of his time is committed. This leaves only $40 - 25$, or 15 hours, for chairs.

Let CHAIRS be the number of chairs the carpenter makes next week. Then the limits on CHAIRS are as follows:

3 CHAIRS ≤ 15 Each chair requires 3 hours and the carpenter has no more than 15 hours for making chairs.

CHAIRS ≤ 4 The carpenter has only got enough timber for 4 chairs.

The inequality 3 CHAIRS ≤ 15 can be manipulated into CHAIRS ≤ 5. Thus the carpenter has time to make up to 5 chairs but his timber is only enough for 4 chairs.

Therefore the maximum number of chairs the carpenter can make next week is 4.

Chapter 5   Introduction to Calculus

Exercises 5.1

1. (a) $6x^5$.
   (b) $1/x^3 = x^{-3}$, so differentiating gives $-3x^{-4}$, or $-3/x^4$.
   (c) $\sqrt{(x^3)} = x^{3/2}$, which differentiated gives $3/2 \; x^{1/2}$ or $1.5\sqrt{x}$.
2. (a) $d(x^2 + 2x^{1/2})/dx = 2x + x^{-1/2} = 2x + 1/\sqrt{x}$
   (b) $d(3x(x-2) + x^{3/2})/dx = d(3x^2 - 6x + x^{3/2})/dx = 6x - 6 + 3x^{1/2}/2$
3. Multiply out the brackets first to get $y = x^2 - 5x + 6$. Then $dy/dx = 2x - 5$.
4. $dv/dt = 8$ – there is nothing special about the notations x and y.
5. The derivative dy/dx of $y = 75x - x^3$ evaluated at $x = 2$ gives the required slope: $dy/dx = 75 - 3x^2$ and so when $x = 2$, $dy/dx = 75 - 12 = 63$. The slope is 63. (For a *marginal* (very small) increase in x above 2, y increases by 63 times the increase in x.)
6. $dy/dx = 2x - 4$. So when $dy/dx = 0$, $2x - 4 = 0$, giving $x = 2$. This tells us that the slope of the graph $y = x^2 - 4x + 2$ is zero at $x = 2$ – in other words, the parabolic curve represented by $y = x^2 - 4x + 2$ is neither increasing nor decreasing at $x = 2$. (This feature characterises a maximum or a minimum value point of a function.)

7. (a) R = tonnes sold × price.
Tonnes sold = tonnes left after n days = $1000 - 100n$. (10% of 1000 tonnes is 100 tonnes.)
Price = $2500 + 100n$ in £s/ tonne. Hence:

$$R = (1000 - 100n)(2500 + 100n).$$

(b) The rate at which revenue is rising is given by

$$dR/dn = d(2\,500\,000 + 100\,000n - 250\,000n - 10\,000\,n^2)/dn$$
$$= -150\,000 - 20\,000n.$$

The farmer is losing potential revenue daily at the rate of £150 000 this loss rate rising further by £20 000 per day or part thereof that the crop is stored. She had better sell quickly!

Exercises 5.2

1. (a) $d^2(x^2 + 2x^{1/2})/dx^2 = d(2x + x^{-1/2})/dx = 2 - 1/2\sqrt{x^3}$
   (b) $d^2(3x(x - 2) + x^{3/2})/dx^2 = d^2(3x^2 - 6x + x^{3/2})/dx^2$
   $= d(6x - 6 + 3x^{1/2}/2)/dx = 6 + \dfrac{3}{4\sqrt{x}}.$

2. (a) $d^2(x^3 - 2x + 3)/dx^2 = d(3x^2 - 2)/dx = 6x$. Using this result we have: $d^3(x^3 - 2x + 3)/dx^3 = d(6x)/dx = 6$.
   (b) $d^2(x^{-2}/2 + 4x)/dx^2 = d(-x^{-3} + 4)/dx = 3x^{-4}$. Using this result we have:
   $d^3(x^{-2}/2 + 4x)/dx^3 = d(3x^{-4})/dx = -12x^{-5}$.
   (c) $d^2(3x^2)/dx^2 = d(6x)/dx = 6$. Clearly, $d^3(3x^2)/dx^3 = d(6)/dx = 0$.
3. (a) The speed is the rate of change of distance covered with time. Hence speed = $v(t) = ds/dt = d(60t + 6t^2)/dt = 60 + 12t$ km/hour, as s is in km and t in hours.
   (b) The rate of acceleration/deceleration is the rate at which speed is changing with time. Hence acceleration/deceleration = $a(t) = d(v(t))/dt = d(60 + 12t)/dt = 12$ km/hour per hour as $v(t)$ is in km per hour and t is in hours.

Exercises 5.3

1. (a) $dy/dx = x + 2$. Hence $dy/dx = 0$ for $x = -2$; $d^2y/dx^2 = 1 > 0$. There is a local minimum at $x = -2$.
   (b) $dy/dx = 3x^2 - 12x + 12 = 3(x^2 - 4x + 4) = 3(x - 2)^2$. Hence $dy/dx = 0$ for $x = 2$; $d^2y/dx^2 = 6x - 12$; $d^2y/dx^2 = 0$ for $x = 2$. This does not help us to decide whether we have a maximum or minimum at $x = 2$. We need to check what happens to the rate of change of the function around $x = 2$. As $dy/dx = 3(x - 2)^2$ its value is never

negative whatever the value of x. So the rate of change of the function dy/dx does not change sign as x increases from just under 2 to just over 2. Hence there is a point of inflexion (neither maximum nor minimum) at $x = 2$.

   (c) $dy/dx = 3x^2 - 12x$; $dy/dx = 0$ for $x = 0$, or $x = 4$; $d^2y/dx^2 = 6x - 12$.
When $x = 0$, $d^2y/dx^2 = -12 < 0$. So there is a local maximum at $x = 0$.
When $x = 4$, $d^2y/dx^2 = 12 > 0$. So there is a local minimum at $x = 4$.

2. (a) Clearly, the cost per seat booked is lowest when the largest possible number of seats is booked on a flight. Hence price per seat is minimum when all 300 seats on the flight are taken by the agent. The price per seat would be £50.

   (b) Let R = total revenue from the agent in £s, and n = no. of seats booked by the agent.
If $n \leq 50$, then $R = 300\,n$. The maximum revenue in this case is when $n = 50$ so $R = £15\,000$.
If $n \geq 50$, $R = (300 - (n - 50))\,n = (350 - n)\,n$.
To find the maximum R we need to find the value of n which makes $dR/dn = 0$. $dR/dn = d(350n - n^2)/dn = 350 - 2n$; $dR/dn = 0$ when $n = 175$; $d^2R/dn^2 = -2 < 0$. Therefore R is maximum for $n = 175$. At that stage $R = 175 \times 175 = £30\,625$. This exceeds the £15 000 revenue obtained without offering a discount to the agent, hence the airline should sell 175 seats to the agent to maximise its revenue from the agent.

3. If an agent has already booked 80 seats, the marginal revenue from the next seat to the airline is found by evaluating dR/dn at $n = 80$; $dR/dn = 350 - 2n$. Hence $dR/dn = £190$ when $n = 80$. The airline should offer the seat to the travel agent.

4. (a) To find the production level t at which $C(t)$ is a minimum we need to compute dC/dt; $dC/dt = t^2 - 12t + 32$; $dC/dt = 0$ for $t = 4$, or $t = 8$; $d^2C/dt^2 = 2t - 12$. When $t = 4$, $d^2C/dt^2 = -4 < 0$ and so $C(t)$ has a local maximum at $t = 4$. When $t = 8$, $d^2C/dt^2 = 4 > 0$ and so $C(t)$ has a local minimum at $t = 8$. To find the absolute minimum of $C(t)$ in the range $t = 0.1$ to $t = 8$ we compute the value of $C(t)$ for $t = 0.1$, $t = 8$, and $t = 10$. We have:

       $C(0.1) = £28\,140$;
       $C(8) = £67\,667$;
       $C(10) = £78\,333$.

Clearly, daily cost is minimum at the lowest possible level of production, 0.1 tonnes per day.

   (b) To find the level of production which makes daily cost maximum we need to evaluate $C(t)$ at $t = 0.1$, $t = 4$ (the local maximum point) and $t = 10$. We already have $C(0.1) = £28\,140$ and $C(10) = £78\,333$. $C(4) = £78\,333$. Hence daily cost is maximum at two

different levels of production, $t = 4$ tonnes per day and $t = 10$ tonnes per day.
(c) The marginal cost of production at $t = 5$ tonnes per day is given by dC/dt evaluated for $t = 5$. We have $dC/dt = -3$ when $t = 5$. Marginal cost of production is negative, which means that when the production level rises marginally above 5, total costs of production fall.

Exercises 5.4

1. (a) $x^2 + K$;
   (b) $\dfrac{-1}{4t^4} + K$;
   (c) $x^3/3 + 5x^2/2 + 6x + K$;
   (d) $5\ln x + K$.
2. (a) $-2/\sqrt{x} + K$;
   (b) $2z^{5/2}/5 + K$;
   (c) $\int (x^2 + 2x + 1)dx = x^3/3 + x^2 + x + K$.
3. Let TC be the total costs of production in £s.
   Then $TC = \int C dn = \int (20 - 0.01n)\, dn = 20n - 0.005n^2 + K$, where K is the constant of integration.
   But if $n = 0$, $TC = £1000$, so
   $TC = 20n - 0.005n^2 + 1000$.
4. The slope of the curve of a function $y = f(x)$ is given by the derivative of the function. So the function we are looking for has derivative $dy/dx = 1/x^2$, or $dy/dx = x^{-2}$. Hence y would be the integral of x^{-2} with respect to x, or

   $$y = \int x^{-2}\, dx = -x^{-1} + K.$$

   We are told that the curve passes through $(x = 1, y = 0)$. Thus we have $-1^{-1} + K = 0$, or $-1 + K = 0$. This gives $K = 1$. Thus the equation of the curve is:

   $$y = 1 - x^{-1} = 1 - 1/x.$$

Chapter 6 Solving Practical Problems with Mathematics

Exercises 6.1

1. We want to find the sum to be invested now at 11.5% per annum, to give a total of £8000 after 5 years. That means that in the notation of the compound interest formula (6.1) developed in this chapter, we want to find C. We have $t = 5$, and $r = 11.5$. So we can say:

   $C(1 + 11.5/100)^5 = 8000$.

Thus $C \times 1.115^5 = 8000$, so that $C = 8000 \times 1.115^{-5}$.

Working this out using the x^y key of a calculator gives $C = £4642.11$ to the nearest penny. You can check that the compound interest formula with $C = 4642.11$ does indeed give an amount of 8000 when $t = 5$ and $r = 11.5$.

In arriving at this result we have had to assume that the interest rate of 11.5% will remain constant – not a very likely assumption over a five-year timescale – and that the purchase will take place exactly at the end of the fifth year. We say that the *present value* of £8000 payable in five years time is £4642.11 at a *discount rate* of 11.5%.

2. As before, let q denote the quantity in each order. Then to meet a demand for D items per year, we will need to place D/q orders, at a total cost of $D/q \times £R$, or $£DR/q$.

The average stock held (with the assumptions of steady consumption and instant replenishment which we made in section 6.5) will be $q/2$, at an annual cost of $q/2 \times £H$, or $£qH/2$.

So the total inventory cost for the year, $£C$, is given by:

$$C = DR/q + qH/2.$$

To find where this has a maximum or minimum, we compute:

$$dC/dq = -DR/q^2 + H/2$$

(remember that the D, R and H are treated as constants in this differentiation).

We then set this equal to zero:

$$-DR/q^2 + H/2 = 0,$$

which can be solved to give:

$$q = +/- \sqrt{(2DR/H)}.$$

You can verify that $d^2C/dq^2 = 2DR/q^3$, which is positive, indicating a minimum, when q is positive (the negative root makes no sense from a practical point of view, since we cannot have a negative order quantity). Thus an order size of:

$$q = + \sqrt{(2DR/H)}$$

gives the minimum-cost inventory policy. This is the well-known *economic batch quantity* (EBQ) or *economic order quantity* (EOQ) formula of operations management theory. You can check that if you substitute $D = 12\,000$, $R = 40$ and $H = £0.24$ into this formula, you get the value $q = 2000$ which we obtained in section 6.5.

3. You need to be careful with units in this question – the stock-holding cost is given in pence per year, but the demand is monthly, and the

order cost is in £s. If we standardise on £s and years as our units, then in the notation of the previous question we have $D = 200 \times 12 = 2400$ units per year, $R = £4$, and $H = £0.12$ per year. Substitution in the formula then gives $q = 400$ units. This means that orders will have to be placed every two months.

(b) Our estimate of order costs was out by 50p, so we have overestimated the true cost of £3.50 by $0.5/3.5 = 14.3\%$ – quite a serious error. We therefore wrongly decided to order 400 items at a time. The annual inventory cost for this policy would be:

no. of orders $\times$ order cost + average stock held $\times$ SH cost
$= 6 \times 3.5 + 200 \times 0.12 = 21 + 24 = £45.$

If we had got the cost right, the true value of q would be given by $\sqrt{(2 \times 2400 \times 3.5/0.12)}$ = approximately 375 units. With this size of order, there will be 2400/375, or 6.4, orders in a year, so that the annual inventory cost falls to:

$6.4 \times 3.5 + 187.5 \times 0.12 = £44.90.$

So although we made an error of nearly 15% in our estimate of the order cost, we are only paying 10p extra as a result – an error of less than 0.3%. This is good news, suggesting as it does that the outcome of our calculation is very stable to even quite large errors in the estimated figures.

Here we have been investigating the sensitivity of the solution to changes in the quantities involved. This kind of *sensitivity analysis* forms an important part of more advanced mathematical problem-solving methods.

4. (a) Let price charged in pence be p, and annual quantity demanded at this price be q units. Then if the relationship between price and demand is linear, we will have:

$q = ap + b,$

where a and b are constants. There are two ways of finding the values of a and b; we give both.

(i) Since $q = 20\,000$ when $p = 40$, we can say that:

$20\,000 = 40a + b.$

Similarly, since $q = 18\,000$ when $p = 50$,

$18\,000 = 50a + b.$

These equations can be solved simultaneously (see section 4.5) to give $a = -200$ and $b = 28\,000$. So the equation linking demand and price is:

$$q = 28\,000 - 200p.$$

The fact that a, the slope, is negative should come as no surprise – demand will decrease as price increases.

(ii) Alternatively, we can say that when price increases by 10 pence, demand falls by 2000 units. So if the relationship is linear, the fall in demand for unit increase in price will be 200 units. But this is simply the slope, or rate of change of demand with respect to price – the a of the linear equation. Thus $a = -200$.

To find b, we note that when $p = 40$, $q = 20\,000$. So, $20\,000 = 40a + b = -200 \times 40 + b$, using the value of a we just found. Solving this linear equation for b gives $b = 28\,000$ as before.

(b) Using the equation $q = -200p + 28\,000$ we have just derived, we want to find the value of p when $q = 0$. Thus we need to solve $0 = -200p + 28\,000$, which gives $p = 140$ pence, or £1.40. However, it is unlikely that a firm would ever increase its prices to the level where demand fell to zero. It is also likely that for any commodity there will be a small market who will purchase the commodity whatever its price, so that the true relationship is unlikely to be linear over the full range of values of p – the picture will probably be more like that shown in the graph below. The precise point at which the linear relationship breaks down will, of course, depend on the nature of the commodity in question. Using the terminology of section 3.1, we could say that the domain of the linear function is unlikely to include $p = £1.40$.

Realistic demand curve

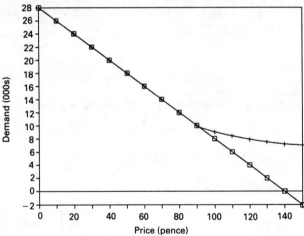

Key: ☐ Linear relationship
+ True relationship

(c) As already noted in (a) above, the value of q when $p = 0$ is given by b, and is 28 000. This represents a ceiling above which demand for the product will apparently not rise even if it is free – however, the validity of the linear equation in this region must again be questioned, as the domain may not include $p = 0$.

5. The graph below shows order value on the horizontal axis against p&p charge on the vertical.

Graph of P&P charges

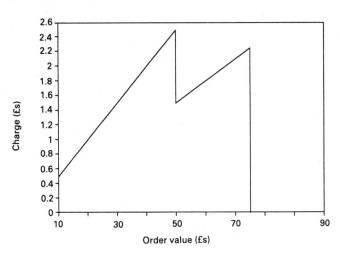

Chapter 7 Simple Statistics

Exercises 7.1

1. | Mark (%) | No. of students |
   |---|---|
   | < 40 | 3 |
   | < 50 | 11 |
   | < 60 | 22 |
   | < 70 | 47 |
   | < 100 | 56 |

2. | | Last year | | This year | |
   |---|---|---|---|---|
   | First | 3 | (12%) | 1 | (4%) |
   | 2.1 | 12 | (48%) | 18 | (67%) |
   | 2.2 | 8 | (32%) | 8 | (30%) |
   | 3rd | 1 | (4%) | 0 | (0%) |
   | Pass | 1 | (4%) | 0 | (0%) |
   | Total | 25 | (100%) | 27 | (100%) |

(Note that the percentages in the last column do not total to 100%, due to rounding.)

3. You could use the layouts:

	North		Sales regions Midlands		South	
	High St	Prec.	High St	Prec.	High St	Prec.
Profitable						
Not profitable						

or

		Profitable	Not profitable
High St	N		
	M		
	S		
Precinct	N		
	M		
	S		

There are several other possibilities, but note how in every case repetition of some categories is needed.

Exercises 7.2

1. The histogram and ogive are shown on page 243. In order to simplify plotting by making all classes the same width, we have assumed that no student obtained less than 30 or more than 80 marks. We can read off from the ogive that 49 students passed the exam (this, of course, assumes an even spread of marks within each band).
2. For this discrete data a bar-chart as shown on page 244 is adequate; percentages rather than actual frequencies have been plotted for ease of comparison. Pie-charts are not a good idea because of the difference in totals between the two years.

Exercises 7.3

1. The mean and s.d. are respectively 9, 4.06 and 90, 40.6.
   The standard deviation here was calculated using the σ_{n-1} key on a

Histogram of exam data

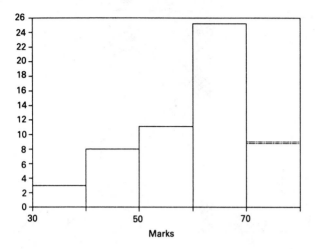

Ogive of exam data

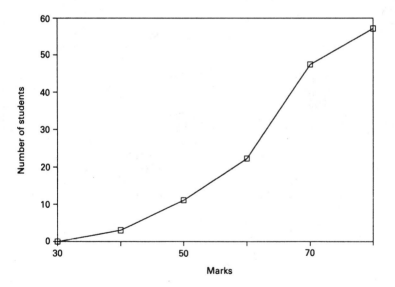

Bar chart of degree data

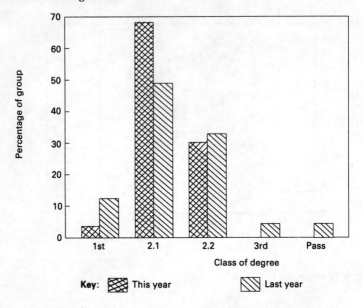

calculator, since the sets of data were described as samples; if you use σ_n instead you will get 3.63 and 36.3 – quite a difference, since n here is only 5.

All the statistical measures for the second set of figures are ten times those for the first set, since the individual data values are ten times as large. This can be useful in simplifying calculations.

2. Assume that lowest mark is 0 and highest is 100.

Mean = 60.98, s.d. = 15.28 (with σ_{n-1}) or 15.41 (with σ_n). There is not much difference here between σ_{n-1} and σ_n because n is quite large at 56. Median is 62.4 (using 28th value rather than 28.5th on the grounds that 56 is a large enough number for the difference to be trivial). Quartiles are 52.7 and 68 (i.e. 14th and 42nd values).

Both the fact that the mean is lower than the median, and the greater distance from Q1 to the median than from the median to Q3, indicate that there is a longer 'tail' on the lower end of the distribution here – something you can confirm by looking back at the original frequency table.

Appendix 1

Exercises A1

1. (a) The area required is shaded on the graph of the function given below. The area is:
$$\int_0^3 2x + 1\,dx = [x^2 + x]_0^3 = 12 \text{ square units.}$$

Graph of $f(x) = 2x + 1$, $0 \leq x \leq 4$

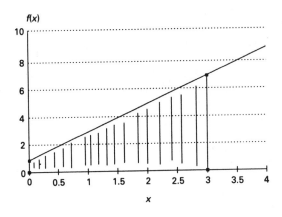

(b) The area required is shaded on the graph of the function given in graph (a) on page 246. The area is:
$$\int_2^4 x^2 - 2x\,dx = [x^3/3 - x^2]_2^4 = 20/3 \text{ square units.}$$

(c) If we plot the function given in graph (b) on page 246 we see that the curve goes below the x-axis for some x-values within the range $0 \leq x \leq 3$.

The definite integral giving the area under BC will have a negative sign. The negative sign should be discarded and the value added to the areas OAB and CED.

To compute the three shaded areas we need to locate the points B and C.

These are where $f(x) = 0$, or $x^2 - 3x + 2 = 0$. Solving this quadratic equation we get the roots $x = 1$ and $x = 2$ which correspond to B and C respectively.

Thus the area required is:
$$\int_0^1 x^2 - 3x + 2\,dx - \int_1^2 x^2 - 3x + 2\,dx + \int_2^3 x^2 - 3x + 2\,dx$$
$$= [x^3/3 - 3x^2/2 + 2x]_0^1 - [x^3/3 - 3x^2/2 + 2x]_1^2 + [x^3/3 - 3x^2/2 + 2x]_2^3$$

(a) Graph of $f(x) = x^2 - 2x,\ 0 \leq x \leq 4$

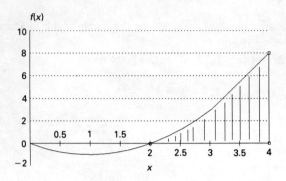

(b) Graph of $f(x) = x^2 - 3x + 2,\ 0 \leq x \leq 4$

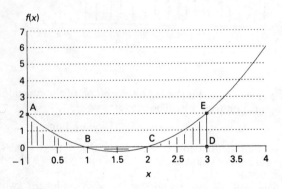

Note that the area under BC is being subtracted from the other two areas so that the negative sign of the corresponding definite integral is eliminated. Thus we have:

Area = $[x^3/3 - 3x^2/2 + 2x]_0^1 - [x^3/3 - 3x^2/2 + 2x]_1^2$
$+ [x^3/3 - 3x^2/2 + 2x]_2^3 = 5/6 + 1/6 + 5/6 = 11/6$ square units.

2. The press saves money up to the tenth year of its life when it begins to cost money. The accumulated savings after x years will be

$\int_0^x (0.5 - 0.05t)dt = [0.5t - 0.025 t^2]_0^x = 0.5x - 0.025x^2$. The press will have paid for itself when the savings equal £0.5 or $0.5x - 0.025x^2 = 0.5$. The solution to this quadratic equation gives $x = 1.06$ and $x = 18.94$. The second root is rejected as it lies beyond the time the press offers any savings. Hence the press will have paid for itself after just over one year.

3. The time in hours to produce 2000 radios is given by $\int_0^{2000} (2x^{-0.1}) dx = [2.222 x^{0.9}]_0^{2000} = 2.222 \times 2000^{0.9}$. Using the x^y function of your calculator you should find that $2000^{0.9} = 935.25$. Hence the total time to produce 2000 radios is $2.222 \times 935.25 = 2078$ hours to the nearest integer.

(You can also use logarithms to get the above answer if your calculator does not have the x^y function. If $y = 2000^{0.9}$ then $\log y = 0.9 \times \log 2000 = 0.9 \times 3.301 = 2.9707$. Finding the number whose log is 2.9707 (using your calculator, or a table of logarithms) we have $y = 934.76$. Hence the total time to produce 2000 radios is $2.222 \times 934.76 = 2077$ hours to the nearest integer, close enough to the answer we got earlier.)

4. The accumulated revenue when n units have been sold will be $\int_0^t (0.4 + x^{-0.125})dx = 0.4t + t^{0.875}/0.875$ in pounds. The accumulated cost when t units have been sold will be $100\,000 + 0.5\,t$ pounds. Thus accumulated profit in £s after t units are sold will be:

$$P(t) = 0.4\,t + t^{0.875}/0.875 - 100\,000 - 0.5\,t.$$

$P(t)$ will be a maximum when $dP(t)/dt$ is 0 and $d^2P(t)/dt^2 < 0$; $dP(t)/dt = 0.4 + t^{-0.125} - 0.5$.

This is zero when $t^{-0.125} = 0.1$, or $0.1\,t^{+0.125} = 1$. Raising to the power of 8 both sides of the equation we have $0.1^8\,t = 1$. Hence $t = 1/0.1^8 = 100\,000\,000$; $d^2P(t)/dt^2 = -0.125\,t^{-1.125}$. This is negative for all t. So accumulated profit has a local maximum at $t = 100\,000\,000$ units.

The absolute maximum of $P(t)$ is found by computing its value at $t = 0$ and $t = 100\,000\,000$; $P(t = 0) = -100\,000$. (In fact, *zero* if the product is not launched.) The accumulated profit when 100 000 000 have been sold is $P(t) = 0.4\,t + t^{0.875}/0.875 - 100\,000 - 0.5t = 40\,000\,000 + 10^8 \times {}^{0.875}/0.875 - 100\,000 - 50\,000\,000 = 10^7/0.875 - 10\,100\,000 = 11\,428\,571 - 10\,100\,000 = 1\,328\,571$ pounds. This will be the absolute maximum profit over the life of the product. It would not be possible for $P(t)$ to increase after $t = 100\,000\,000 = 10^8$ because marginal revenue would have fallen to $0.4 + 10^{8 \times (-0.125)} = 0.5$, the same as the marginal cost per unit.

Index

absolute value 31
absolute maximum/minimum 145
addition 2
antiderivative 204
area under curve 208
asymptotes 94

bar chart 183
base 39
base of natural logarithms 19
billion 8
brackets, operations with 50

Cartesian plane 61
census 175
common logarithm 86
constant of integration 152
continuous data 275
coordinates 62
cumulative frequency 179

decimals 14
  recurring 18
definite integral 205
demand function 71
denominator 10
  common 11
dependent variable 59
depreciation 166
derivative 135
  higher order 138
difference 7
differentiation 132
discrete data 175
dividend 8
division 4
  involving negative numbers 4
  ways of writing 7
divisor 8
domain (of a function) 60

equations 19
  linear, solution of 99
  quadratic, solution of 104
  simultaneous 114
exponent 39
exponent notation 8
exponential
  function 81
  growth 82

factorial 32
factors 55
fractions, cancelling of 10
  notations for 10
frequency 176
  cumulative 179
  distribution 178
function 59
  discontinuous 94
  exponential 81
  inverse 84
  linear 63
  logarithmic 85
  quadratic 78

gradient 66
graph
  log-log 92
  plotting 61
  semi-log 90

Index

histogram 185
hyperbola 94

indefinite integral 151
independent variable 59
index 39
inequalities 32
  solution of 123
integer 2
integral
  definite 205
  indefinite 151
integration
  formula for 151
intercept 67
inventory 170
inverse function 84

linear function 63
logarithm 86
  common 86
  natural 86
logarithmic function 85
log-log graph 92
lowest common denominator 11

market equilibrium 76
maximum
  absolute 145
  local 78
  to find position of 139
mean 191
median 195
minimum
  absolute 145
  local 78
  to find position of 139
mode 196
modulus 31
multiplication 4
  involving negative numbers 5
natural logarithm 86
number line 2
numerator 10

ogive 189
origin 61

parabola 78

percentage 23
  arithmetic with 26
  calculation of 24
  conversion to decimals 28
  using a calculator 25
pie chart 161
point of inflexion 144
population 175
power 39
  fractional 46
  negative 48
  zero 47
present value 171
priority of operations 5
product 7

quadrant 63
quadratic function 99
quartiles 201
quotient 8

range 202
rate of change 64
root 39
  of an equation 98
rounding 20

sample 175
scientific notation 8
semi-log graph 89
significant figures 21
simultaneous equations 114
sinking fund 162
slope
  definition of 66
  to compute using calculus 134
standard deviation 197
subtraction 2
sum 7
summation sign (sigma) 30
supply function 74

tangent 132
turning point 144

variable, algebraic 35
  statistical 175

zero, division by 6